Numbers Explained

Part One - Essentials

Stephen Miller

This book is for adults or older students who lack ability or confidence in arithmetic and who would like to understand numbers better. It is based on the author's experience as a classroom teacher and as a private tutor.

Published by CompletelyNovel.com

Cover design by Lorraine Payne

Associated web-site by Big Cat Digital Ltd

Printed in England

ISBN 9781849144360

Dedication

When I was a child, I thought that I was no good at maths. At my primary school (this was in the 1950s), I was required to perform calculations like "divide 3 tons 5 hundredweight 1 quarter 6 stone 12 pounds by three". If I got the answer wrong, I would have to do it again, and again and again, until I got it right. I hated it.

At my secondary school I encountered geometry for the first time. By a quirk of time-tabling we had a special teacher who used Euclid's theorems. I loved it. I appreciated the elegance of the proofs and I lavished great care over the constructions. The rest of the maths was a different story, where we had algebra hammered into us by a man with a short fuse, a disdain for ignorant little boys and a caustic tongue. He reinforced my feelings of inadequacy.

I was lucky. My father was a science teacher who was also trained in maths and one day, before I was due to take an exam, he sat me down and just went through the things that I needed to know, making notes on a single sheet of paper. I was amazed. I could see that there wasn't really all that much to remember and that it all made sense.

Still, I persisted with the idea that I was no good at maths and I tended to drift along in lessons. I was in my GCE 'O' level year when my teacher was handing out books and said to me in a quiet aside "If you don't get a grade 1, I shall be very cross." I was so surprised, but it changed my attitude completely. It finally dawned on me that maths wasn't that hard and that, if I wanted to, I could be very good at it. (I did get that grade 1.)

This book is dedicated to the memory of my father, Dennis Miller, who showed me that maths was not so difficult really, and to the memory of my teacher, Jack LeFevre, who gave me confidence.

It is also dedicated to the millions of you out there who think that you can't do maths.

Numbers Explained – Part 1 - Essentials

Contents

In Numbers Explained Part 2 – Further Knowledge

Numbers Explained Part 2 is to be published later in 2014

Chapter 1 – Introduction

Lots of people find maths difficult. I am not talking about advanced maths here; many people are lacking in basic numeracy. There are many well-educated people who would struggle to add up a column of figures.

"I am no good at maths", said with a wry smile, is readily admitted, because the speaker will often find that they are talking to a fellow sufferer, who gladly dropped maths as soon as their school let them. Innumeracy is socially acceptable, unlike illiteracy which is carefully hidden.

This is more than a widespread lack of interest in maths. It is a real dislike of the subject. A conversation at a party: "Oh, you are a teacher. What subject?" "Maths." The questioner recoils. You wouldn't get the same reaction to Geography.

I have been a maths teacher in both primary and secondary schools in England. I have also given private tuition in maths to a wide range of ages and ability. From my experience, I strongly believe that there is nothing inevitable about the difficulties that people have with maths.

There are two reasons why they have given up on maths.

Firstly, there is the step-by-step nature of mathematical knowledge. In some ways, this is a great help, because there is a clear thread of argument that can be followed from beginning to end. I have never encountered anyone who was incapable of following that thread, if they want to. However, it causes much trouble, because you cannot miss any step in that argument.

Who has a perfect attendance record in school? Take two weeks out of school for some reason and perhaps you have missed the geography of Japan. This may be regrettable but it won't affect your learning about other parts of the world. However, you have also missed lessons on fractions and this will undermine all subsequent work on number and algebra. That is, unless you are lucky enough to have someone available who can explain what you have missed.

This catch-up work must take place as soon as possible, otherwise confusion and failure may well poison your attitude to maths. (This is made worse by everyone telling you what an important subject maths is.) In the face of this pressure, it is quite natural to give up and think, "Oh well, I'm just no good at this."

In England and Wales, the National Curriculum tries to deal with this problem by repeating topics, each time with increasing depth. I have some sympathy with this approach, but students who have not understood something the first time need a slower pace than those who have already 'got it'. The resulting lessons will either fail with the first group or bore the second and probably please neither.

The second source of difficulty is the way that maths is often taught. Teachers are under pressure to get results and there is a temptation to teach "tricks for ticks". This means that students are told, "Do this and then do that and then the answer is here" without paying any attention to what is being achieved and how it fits into the larger scheme of things. You are not required to understand it, just know how to do it.

This tendency is most marked in Primary schools, where it is not uncommon for teachers to be a bit shaky at maths themselves. Unfortunately, it is precisely when children are starting out in maths that they most require an informed and sensitive approach to their learning. They need lots of play and discussion and activities that involve number and shapes and relationships between objects. If they are required to engage in formal, abstract work on paper at too early an age, they are likely to be put off maths for life. This often happens.

A good example of inappropriate early maths is a mad insistence that children should learn multiplication tables when they are far too young and may have only a dim grasp of the nature of multiplication. True, at some point we all should knuckle down and learn these useful number bonds. It isn't really difficult to do this provided (a) you can see the point of it and (b) you haven't discovered that you are a failure because you just can't remember that it's six nines that are fifty-four and seven eights that are fifty-six. And all this when you have no real idea what multiplication is because you have only just learnt to count and write numbers.

In England and Wales, at both primary and secondary level, the tendency to teach procedures rather than understanding is reinforced by school league tables, SATs (Standard Assessment Tests) and the regular testing that now accompanies the National Curriculum. The result has been a focus on getting through the next Key

Stage test at all costs. At its worst, the approach of the students is to just mug up what they need to do to get through the next test and then forget it – a strategy encouraged by the modular exams that were introduced for GCSE and are now (2014), at last, to be abandoned. Good. It is truly depressing to be asked "Are we going to be tested on this, sir?" Then, after replying, "No, but it is interesting", see their eyes glaze over.

It is true that many students succeed in maths despite the problems that I have outlined. They have early experiences that enable them to infer patterns in the jumble of strange tasks that they encounter in maths lessons and somehow they extract meaning from it. They score sufficient ticks for the tricks that they perform to retain a positive attitude and they can see the relevance to their everyday world. However, far too many people struggle to remember what they have to do, can't see the point to it and become profoundly alienated.

If you are one of the majority who either saw maths as a hoop that you just had to jump through to get on in life or saw maths as an ordeal to be endured, **do not despair**. As a private tutor, often called in with a few weeks to go before an important exam, I have invariably found that I have had to go right back to the start, examining topics that the student first encountered as a small child, sorting out misconceptions and filling in gaps in understanding. Then, what was a buzzing mish-mash of unrelated facts to remember and regurgitate on demand, suddenly collapses down into an orderly set of quite simple ideas that make sense. The subject becomes manageable and even interesting, and best of all, the student can do it. I call it the "Gee, I am a swan!" experience.

I have found myself explaining the same ideas over and over again. This book attempts to set down those ideas. It may not suit everyone and it is a poor substitute for hearing those ideas explained by someone sat next to you who can transmit their understanding and enthusiasm through their tone of voice and who can answer any questions. It is, however, a lot cheaper and you can proceed at your own pace.

It concentrates purely on number work, because a facility with numbers is most relevant to everyday life and because most branches of maths rely on number to some extent. I hope you find it at least useful and, at best, life-changing.

Finally, I'd like to tell you a story.

I was studying at university and thought that I would do some private tuition to help keep the wolf from the door. I had a reply from a man who rather hesitantly told me that he didn't want tuition for his children, but for himself. "Really?" I thought. I said I'd come and see him and discuss what he wanted.

He was in his forties and worked as a tool-setter in an engineering works. He had left school with no formal qualification in maths. He said that he wanted to catch up what he had missed when he was at school. Anyone reading this who is still at school should take note: this is a very common theme. I'd like a pound for everyone who has told me "I wish I could have my time in school again. I wouldn't mess around this time." Well, as I say, I have heard this sort of thing before, but here was a man who actually proposed to do something about it.

He told me, "When I am at work, the engineers come round and do calculations to set up the machines correctly. I want to know what they are doing and I want to be able to do it." "OK', I said, "I've never done anything like this before, but let's see what we can do."

He was wonderful! He was the best student I have ever had, either private or in school. He just soaked up maths like a sponge. I only did about ten one-hour weekly sessions with him, although I did set him 'homework', which he did assiduously. We started with some simple set theory to get started with the different sets of numbers that we use in arithmetic. We moved quickly on to integer arithmetic and the importance of place value. Then we did fractions and then on to decimals. Finally, just because he wanted to do some more maths, we did some simple linear algebra, linking it to graphs and we finished off with some basic simultaneous equations. All in ten weeks!

It is **never** too late to learn.

Chapter 2 – How to Use This Book

The book is split into two parts. Part 1 – Essentials covers everything that you are likely to encounter in everyday life. Part 2 – Further Knowledge covers topics that are probably only necessary if you want to tackle maths in the context of science and engineering.

It is very unlikely that you have no knowledge at all of maths, so this book will often tell you things that you already know. You may find that encouraging.

However, it will also contain a great many things that you don't know. Some of these will be maths topics that have previously caused you trouble. There will be a temptation to skim through the book to pick out these topics. Please don't do this, because there will almost certainly be other things that you don't understand because you have not encountered them before. These are 'unknown unknowns', to quote the former US Secretary of State for Defense, Donald Rumsfeld.

To make the point further, this book has been proof-read by, among others, my wife. She is no slouch at maths, as she has a PhD in physics. All the same, she went through the same 'tricks for ticks' regime as most of us. Consequently, she found things in this book that she didn't know, and she discovered the reasons for other things that she had previously just taken on trust.

Maths is built in logical steps. Each new idea depends upon previous ones. So, please read through Part 1, from the start, without skipping anything unless it says that you can, at least once. This is most important.

You will be able to read some chapters very quickly. Others may take longer. Make sure that you have read, understood and thoroughly digested each chapter before moving on to the next.

Some people's experience of maths has been so bad that their minds tend to go blank when they see numbers. They can even experience panic attacks. If you are one of them, it is particularly important that you take your time.

A lot of maths consists of learning the meanings of words. Some are special mathematical words, but others are everyday words that also have a special mathematical meaning. There is a glossary at the end of the book that may help you if you see a mathematical term that you don't know.

> **Important points to remember are shown in bold in boxes like this. There is a list of them at the end of the book.**

> Anything shown like this in a shaded box can be skipped, but throws further light on the surrounding text. You don't need to remember it, but it may be of interest.

If you encounter something that you cannot understand after several attempts, please get in touch with me so that I can help you and so that I can consider revising the book. Please use the contact form on the web-site http://numbersexplained.co.uk.

If you want to practice any of the arithmetic procedures that are explained in this book, you can also download suitable exercises from http://numbersexplained.co.uk. All questions are provided with worked answers so you can see exactly how they should be done. Alas, there are no prizes.

One final note: this is a first edition. It has been carefully checked by experienced editors, but it is possible that there may be errors that have escaped the net. It is amazing how mistakes can creep in undetected. So, if you find something that is obviously wrong, or something that is puzzling you and may be due to an error, please let me know. Then I can put a note on the web-site for other readers and I can eventually sort the problem out in a later edition.

Chapter 3 – Counting

You may be thinking "This is ridiculous! I don't need to be taught how to count."

I'm sure you don't, but this chapter is about counting, partly because it is the start of all that we are going to explore as we look at more exotic forms of numbers. However, the main reason is that I want you to know that you have already mastered some quite complex ideas.

I have a favourite way of showing that there are some rules that underpin counting that everybody knows without realising it. I deliberately break the rules.

Firstly, here are some cabbages.

I would be amazed if you couldn't tell me that there are five cabbages.

Now I am going to demonstrate that there aren't five but four. Next to each cabbage I will write down the counting numbers as if I were saying them out loud.

one   two   three   four

As you can see, there are four.

At this point, the person I am demonstrating to gives me a funny look and says, "No, that's not right. There are five." And I say "What did I do wrong?" "You missed one out." So there is a rule that every cabbage has to have its own number.

There are actually a lot of rules governing counting.

Here are some more deliberate errors in counting. What rule is being broken in each case?

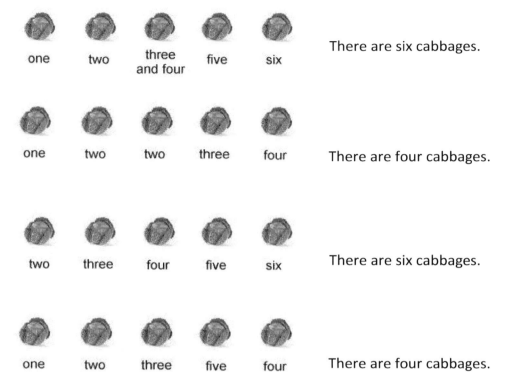

one   two   three and four   five   six
There are six cabbages.

one   two   two   three   four
There are four cabbages.

two   three   four   five   six
There are six cabbages.

one   two   three   five   four
There are four cabbages.

The broken rules were:

1) You can only give each cabbage one number
2) You can only use each number once
3) You have to start counting with one
4) Counting numbers are in strict order

Most people have learnt these rules and apply them without error by about the age of seven. Yet if you were to ask them, "What are the rules of counting?", they would look blank.

Most mathematical ideas are simpler than counting. If you can already count, then you can learn to handle the rest of everyday maths, if you want to. Mostly, it involves learning the precise meaning of various words and symbols.

Sets

A very basic idea in maths is that of a **set**. I am going to use this word quite a bit, so I am going to define a set as a collection of things.

A set is a collection of things.

(If you see something in a box like this, you need to remember it. There is a list of these things to remember at the end of the book.)

We need to consider a special type of set – an **ordered** set. This is another fairly simple idea: the members of the set can be put into a meaningful order.

I am now going to use this idea of an ordered set to examine more closely what is going on in counting.

The Set of Counting Numbers

Suppose you are going to lay the table for dinner. For everyone at the table there will need to be a chair, a knife, a fork, a spoon, a plate, and so on.

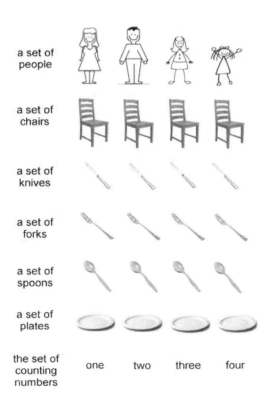

a set of people

a set of chairs

a set of knives

a set of forks

a set of spoons

a set of plates

the set of counting numbers   one   two   three   four

The last set in the diagram is taken from the ordered set of **counting numbers**, and the last member in this set is four. So, there are four objects in each set. Knowing this is very useful when laying the table.

For this meal for four, we can pretty well see the situation at a glance and we might be able to lay the table without counting. However, suppose it were a banquet for four hundred people. Then we would almost certainly have to count out four hundred chairs, knives, and so on to make sure that we had enough for everyone with none left over.

| This relationship between sets is called one-to-one correspondence. | (Boxes like this contain things of interest that are not essential to know.) |

It is said that there are (or were) primitive societies whose counting numbers went "one, two, a lot". We certainly couldn't get by in our complex world without the ability to count.

I will now define exactly what is going on in counting.

> **The counting numbers are an ordered set of labels that we assign, in order, to every member of a set of objects. The last label that we use is the number of objects in that set.**

This ordered set of labels, the counting numbers, and the procedure for counting, are so important in a modern society, that virtually everybody learns how to do it at quite an early age. It is so taken for granted that asking someone "Can you count?" is much the same as asking "Are you stupid?". Even people who declare themselves to be "no good at maths" can usually count.

How many labels do we need? Obviously, we might need to count very large sets of objects, but we can't have a unique label for every number or the task of learning them would become impossible. Instead, we have an ingenious method of re-using a fairly small number of labels – usually ten: 0 1 2 3 4 5 6 7 8 9. This trick – place value – will be the subject of the next chapter.

Numbers Explained

As I hope you can see, counting is not as simple as it seems. If you can count, you have already mastered some important mathematical ideas and a detailed procedure. There is no reason why you shouldn't learn any of the remaining ideas and procedures that follow in this book.

Finally, here is an observation about counting and small children. It is not necessary for you to know this, but it is quite remarkable and is worth a quick read.

The Concept of Number in Small Children

You can try this classic, startling experiment yourself if you can get the attention of a child who can talk – a four year-old is about right. Show them two identical lines of counters (or sweets or cakes or something that interests them) and say, "This line's yours and this line is mine. Who has most, you or me, or have we got the same?" The child will say "The same." Now spread your line of counters out so that it's the same number of counters making a longer line. Do this in full sight of the child. Now repeat the question. The child will now say "You have."

It is hard to believe that this is true, but if you try it you will see for yourself. You can get similarly baffling results by pouring liquids into different shaped glasses and asking small children about it.

The Swiss developmental psychologist Jean Piaget conducted experiments like this on small children and came to the conclusion that childrens' minds are different to those of adults.

Nearly all adults have the concept of the number of a set of objects and the idea that it remains the same even if you change other properties of the set such as where each member is located. This concept does not form in the minds of young children for some time. The exact age varies, but it is usually there by about seven and the change is probably the result of repeated observation as the child plays with objects.

Anyone who thinks that small children should stop playing so much and learn mathematical facts by rote at an early age should try this experiment and reflect on it. I can recommend it to Secretaries of State for Education, in particular.

Chapter 4 – Place Value

We can express numbers in words – one, two, three, and so on – or in figures. When we write numbers in figures, we use only ten characters:

0 1 2 3 4 5 6 7 8 9

If we want to write any numbers larger than nine, we have to organise the figures into columns. Each column has a **place value**. The easiest way to see how this works is to start counting and see what happens. The first column has a place value of **units**. This means we are counting things one at a time. We will head the column with a U to show this.

U	
0	Zero or nought
1	One
2	Two
3	Three
4	Four
5	Five
6	Six
7	Seven
8	Eight
9	Nine

Now we have run out of characters. We start a column on the left in which we are counting in groups of ten and we will head the column with a T to show this. A 1 goes in this column to represent ten and a zero goes in the units column.

T	U		
1	0	Ten	(one ten and no units)
1	1	Eleven	(one ten and one unit)
1	2	Twelve	(one ten and two units)
1	3	Thirteen	(one ten and three units)
1	4	Fourteen	(one ten and four units)
1	5	Fifteen	(one ten and five units)
1	6	Sixteen	(one ten and six units)
1	7	Seventeen	(one ten and seven units)
1	8	Eighteen	(one ten and eight units)
1	9	Nineteen	(one ten and nine units)

Now we have used up all the characters again in the units place value, but that's not a problem. Twenty is two tens, so we put a two in the tens place value and a zero in the units, and continue thus:

T	U		
2	0	Twenty	(two tens and no units)
2	1	Twenty one	(two tens and one unit)
2	2	Twenty two	(two tens and two units)
2	3	Twenty three	(two tens and three units)
2	4	Twenty four	(two tens and four units)
2	5	Twenty five	(two tens and five units)
2	6	Twenty six	(two tens and six units)
2	7	Twenty seven	(two tens and seven units)
2	8	Twenty eight	(two tens and eight units)
2	9	Twenty nine	(two tens and nine units)

This pattern is repeated for thirty, forty, and so on, until we get to ninety nine.

T	U		
3	0	Thirty	(three tens and no units)
3	1	Thirty one	(three tens and one unit)
.	.	.	.
.	.	.	.
3	9	Thirty nine	(three tens and nine units)
4	0	Forty	(four tens and no units)
4	1	Forty one	(four tens and one unit)
.	.	.	.
.	.	.	.
9	7	Ninety seven	(nine tens and seven units)
9	8	Ninety eight	(nine tens and eight units)
9	9	Ninety nine	(nine tens and nine units)

At this point we have used up all the characters from 0 to 9 in both the units and the tens place values. To continue, we start another place value for hundreds and then fill up the units and tens all over again.

H	T	U	
1	0	0	One hundred
1	0	1	One hundred and one
1	0	2	One hundred and two
1	0	3	One hundred and three
.	.	.	.

.	.	.	.	
1	0	9	One hundred and nine
1	1	0	One hundred and ten
.	.	.	.	
.	.	.	.	
1	9	7	One hundred and ninety seven
1	9	8	One hundred and ninety eight
1	9	9	One hundred and ninety nine
2	0	0	Two hundred

This pattern keeps repeating.

The beauty of place value is that we need never stop. Whenever we have used up all the characters from 0 to 9 in all the place values, we simply start another one to the left. So, when we get to 999, we start a **thousands** place value to the left and write 1 000, 1 001, 1 002 and so on. When all four columns are used up as we get to 9 999, we start a **tens of thousands** place value to the left and carry on with: 10 000, 10 001, 10 002 and so on.

Note that every column is worth ten times the one on its right.

Note also the importance of zero. When we write numbers normally, we do not have H T U heading the place values. It is only because zeros fill up empty place values that we can distinguish between 2, 20 and 200 or between 21 and 201.

Every place value is worth ten times the one on its right.

Zeroes are used to fill up empty place values.

Notice also that the words that we use to represent numbers are also loosely based on groups of ten. In English, we have two special words 'eleven' and 'twelve', and then words ending in '–teen' to take us up to twenty. Then the tens are all words ending in '-ty'. After that it is all regular: hundreds, thousands, etc.. If you speak another language, you might like to consider how its counting words work.

Modern usage is to put a small space between each group of three columns e.g. 9 254 instead of a comma e.g. 9,254 because a comma is a decimal point in some countries.

The earliest known example of counting consists of scratched marks on a piece of bone dated some 50 000 years ago. Making a mark for each item that you are counting is called a tally system, although tallies can be beads or 'counters'. Tallies are often grouped. An example is the 'five bar gate' system: 1, 11, 111, 1111, ~~1111~~, ~~1111~~ 1, ~~1111~~ 11 and so on.

Counting systems that use place value have replaced older tally methods of representing numbers because these only work for relatively small numbers and rapidly become clumsy and unreadable.

The Roman numeral system: I, II, III, IV, V, VI, VII, VIII, IX, X, XI and so on, attempts to cure this problem by introducing special characters X for ten, L for fifty, C for one hundred, D for five hundred and M for one thousand. This is moving in the direction of a proper place value system but it fails because the place values do not increase regularly and they stop at one thousand. Also, it uses the convention 'one before' to represent four as IV and nine as IX, which causes confusion.

All such counting systems fail to handle large numbers in a readily readable form and are also quite impossible to use in calculations.

Consider what a clever idea place value is. We can write numbers that are as large as we like but we use only ten characters from 0 to 9.

Number Bases

The words that we use to count up to a hundred have a pattern to them based on the number ten. There are special names for eleven to nineteen and then there are names for all the tens from twenty to ninety. These are our every-day counting numbers. They are a **base ten** number system because there are ten characters and each place value is worth ten of the one on its right.

You may wonder why we use ten as our number base. No one really knows, but it seems likely that our numbers grew out of tally systems (making marks to represent numbers). These were based on counting on our fingers, and we have ten digits on our hands. It is interesting to note that each place value in a number is often referred to as a **digit**, so that a number in hundreds is a '3-digit number'. It is possible to use a different number base and we look at this in Part 2.

It is known that in the 3rd Century BC in Sumeria (roughly modern Iraq), they used a base sixty system (also called sexagesimal), although an early version did not have a symbol for zero and numbers larger than sixty appeared to be understood purely from their context. One of the advantages of base 60 numbers is that a single digit can be divided by 2, 3, 4, 5 , 6, 12, 15, 20 and 30 to get another number, whereas our base ten system digits can only be divided by 2 and 5. A difficulty is the need to have 60 characters.

It is thought that our way of measuring time with 60 seconds to a minute and 60 minutes to an hour is a legacy of the Sumerian system of counting. Probably this is for the convenience of being able to divide easily by 2 and 4 to give half and quarter hours. Also we have 360 degrees in a full turn when measuring angles for the same reason.

In Part 2, we shall look at different number bases in more detail. In particular, we shall see how electronic devices use a base 2 system called binary and how information technology engineers represent binary numbers in either base 8 (also called octal) or, more commonly, in base 16, called hexadecimal or 'hex'.

Large Numbers

Very large numbers are written with the digits in groups of three. The diagrams below show how they are organised with the names of each group of three:

Trillions			Billions			Millions			Thousands					
H	T	U	H	T	U	H	T	U	H	T	U	H	T	U
									1	3	2	4	0	9

is 'one hundred and thirty two thousand four hundred and nine'.

Trillions			Billions			Millions			Thousands					
H	T	U	H	T	U	H	T	U	H	T	U	H	T	U
							3	9	0	6	5	7	0	0

is 'thirty nine million, sixty five thousand, seven hundred'. Note the use of zero to fill up the empty hundreds of thousands place value as well as the empty tens and units place values.

Trillions			Billions			Millions			Thousands					
H	T	U	H	T	U	H	T	U	H	T	U	H	T	U
			2	2	5	0	0	0	0	0	0	0	0	0
	7	6	2	0	0	0	0	0	0	0	0	0	0	0

are 'two hundred and twenty five billion' and 'seventy six trillion, two hundred billion'.

Here are two examples without the column headings: 328 420 000 is 'three hundred and twenty eight million, four hundred and twenty thousand'. 350 050 000 000 is 'three hundred and fifty billion, fifty million'.

It is all absolutely regular. Just read each group of three as a HTU number and attach its group name.

(Note: in Britain, we now use the US billion which is 1 000 000 000. We used to call this a 'thousand million' and a billion was 1 000 000 000 000. Some people may still use this old system, so be aware that there is a possibility of misunderstanding.)

It is not easy to understand just how big large numbers are. The names million, billion, trillion seems very similar and we have bogus words like 'squillion' and 'gazillion' that simply convey the idea of a staggeringly large number. If you read that the cost of the Eurotunnel project was £ $4\frac{1}{2}$ billion, what does that mean? It is obviously a lot of money, but how much is a lot?

A good way to appreciate the scale of large numbers is to consider time measured in large numbers of seconds. When you do this you get the following result:

A thousand seconds	=	Nearly 17 minutes
A million seconds	=	About eleven and a half days
A billion seconds	=	Nearly thirty two years
A trillion seconds	=	About thirty two thousand years

So, a billion seconds is getting on for half a human lifetime and a trillion seconds is much longer than the recorded history of mankind.

The set of integers is infinite in size – no matter how large a number you can think of, there is always another larger one.

The idea of infinity is often abused. People might describe the number of grains of sand on the beach as infinite. It isn't. It may be very large, but it is a finite number. In principle, you could put each grain into one-to-one correspondence with integers and thus count them. Finite sets are countable.

However, we can conceive of infinite sets of things. You might assert that there is no limit to time (please feel free to argue about this). In that case, the existence of the universe would be an infinite number of seconds. That doesn't mean that we couldn't, in principle, count them. All we have to do is to put them into one-to-one correspondence with the set of integers, which is also infinite. So it is possible for a set to be infinite and still be countable.

The really weird thing about an infinite set is that it can be put into one-to-one correspondence with a sub-set of itself. For instance, the set of even numbers is a subset of the set of all integers – it is every other number. However, it can be counted – for each counting number, you can just double it to get the corresponding even number. How can that be? Well, you can do this because you never have to stop counting. It would only be impossible if the set were finite and you had to stop at some point.

If all this makes your head hurt, that's quite normal.

Number Lines

A useful way of displaying the set of counting numbers is to set them out along a line, starting from zero and spacing them evenly out like this:

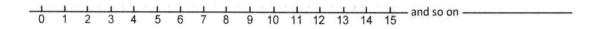

This book uses number lines a lot in order to explore the rules that govern numbers. These rules are called arithmetic.

Chapter 5 – The 'Four Rules' of Number

It is easy to see how numbers are connected to each other. If you put three cups on the table and then later you add another two cups, there will then be five cups on the table. This is a completely reliable phenomenon – you never find three plus two making anything other than five.

Similarly, if you take a cup away from the five cups on the table, there will always be four left.

It doesn't just work with cups, but with all objects – you don't get a different result if you do the same thing with shoes or watering cans. Small children don't know this – why should they? (How about chimpanzees, or parrots?) However, it is all so familiar to adults that they take it for granted.

These connections of numbers are called **number bonds**. We need to know number bonds for the counting numbers up to 9 and it is very helpful to know them for larger numbers. Fortunately, we don't have to learn them for all numbers, because we have a number system that uses place value. Consequently, we can work with large numbers by dealing with each place value separately.

Numbers bonds are the key to performing **calculations**. We do calculations so often that it is easy to forget how remarkable this is. Calculations are abstract models of the real world that we can use to make predictions.

For a simple example, if I want to decorate a room with wallpaper, I could stand in the room and guess how many rolls I need. My chances of being right are not very good. If I make measurements and write the numbers down, then I can perform calculations on these numbers that will enable me to predict accurately the required number of rolls. Calculating usually beats guessing.

More impressively, we can construct a numerical model of the solar system and of the trajectory of a spacecraft so that it can touch down on a remote planet. This would be utterly impossible without calculations.

Knowledge is power.

The most common calculations are addition, subtraction, multiplication and division. These are often called the 'four rules' of number. There are other forms of calculation, but we will consider them in Part 2.

Addition

This is the most basic number bond. The words 'add', 'plus', 'and' and 'increase' and the sign + are all used to show addition. The result of addition is called a 'sum' or a 'total'.

If we represent addition on a number line, it is a movement 'up' the line to the right. This implies that a single number, for example three, is actually an addition that starts from zero, and can be written as +3. Usually the + sign is left out. On a number line, 3 (or +3) looks like this:

We often want to add two numbers, and here on a number line is the addition number bond 'three plus two equals five':

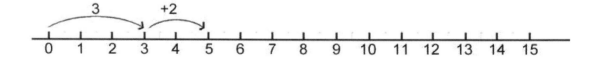

We can add more than one number, for example 'three plus two plus four equals nine':

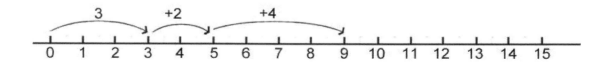

We get the same result if we change the order of numbers that are added together:

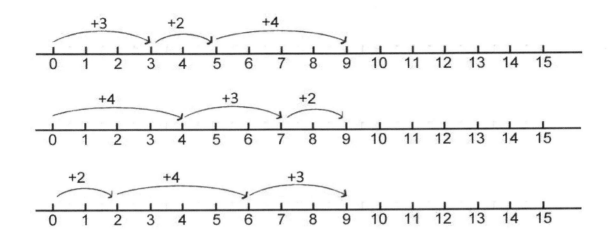

As you would expect, 3 + 4 + 2, 2 + 3 + 4 and 4 + 2 + 3 also give 9.

This means that we can calculate several additions from left to right or from right to left:

From the left:		From the right:
	3 + 5 + 2 + 3		3 + 5 + 2 + 3
=	8 + 2 + 3	=	3 + 5 + 5
=	10 + 3	=	3 + 10
=	13	=	13

As this suggests, you can reverse the order of an addition so that 3 + 4 = 4 + 3 = 7.

Also, if we have a list of numbers to add or 'tot up', we can start at the top or the bottom. In fact, it is useful to tot up a list both ways and, if we get the same answer each time, we can be fairly confident that there is no mistake.

If you would like to know the technical language for the properties of addition, here it is.

Addition is **associative**, which means that it doesn't matter if you work out 3 + 2 + 4 by adding 3 and 2 first to give 5 + 4, answer 9, or by adding the 2 and 4 first to give 3 + 6, answer also 9.

Addition is also **commutative**, which means the mirror image property that 4 + 7 = 7 + 4 = 11.

Also, zero is not just the starting point of all counting numbers, it is also the **identity** member of the set for addition because adding zero has no effect.

Subtraction

Subtraction is shown by words such as 'minus', 'less', 'reduce' and 'take away' and the sign -. The result of subtraction is called a 'difference' between two numbers.

If we represent it on a number line, subtraction is a movement "back down" the line to the left. We can have a single number e.g. −3. We always start counting from zero, so this takes us into the set of negative numbers which is the topic of a later chapter. On a number line, minus three looks like this:

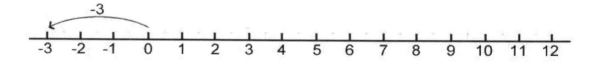

In the last number line diagram, we extended it to the left to provide negative numbers and in the process revealed that zero is a number in its own right, not just an indicator of an empty place-value column. It has to be a number otherwise the step between -1 and 1 would be wrong.

Until we have looked at negative numbers in more detail later, we will avoid them by only subtracting smaller numbers from larger ones, for example, seven minus three, meaning "go 7 steps up the number line from 0 and then 3 back "which looks like this:

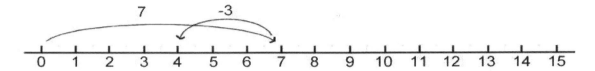

Subtraction is very different from addition in that the order is important. Whereas 7 + 3 = 3 + 7, it is <u>not</u> true that 7 − 3 = 3 − 7. Incidentally, we can use a 'not equals' sign ≠ to show this: 7 − 3 ≠ 3 − 7.

The reason for this is clear if we show both 7 - 3 and 3 − 7 on number lines. One gives 4 and the other gives -4 :

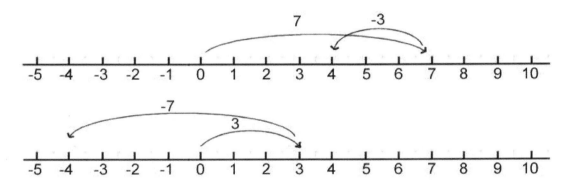

Unlike addition, we can't subtract several numbers in any order and get the same answer. Here is a subtraction involving three numbers performed twice. The first time the subtractions are done from left to right and the second time from right to left:

From the left:		From the right:	
	12 − 7 - 2		12 − 7 - 2
=	5 - 2	=	12 - 5
=	3	=	7

As you can see, you get a different answer depending on which subtraction is done first.

More formally, unlike addition, subtraction is non-associative and non-commutative.

This means that arithmetic expressions that include subtractions can be ambiguous. Ambiguities are resolved by the use of brackets, as explained next.

The Use of Brackets

So, if we want to subtract seven and two from twelve, how do we know which one we mean? The way to sort it out is to use a pair of brackets to show which subtraction comes first, like this:

$(12 - 7) - 2$ $12 - (7 - 2)$
=   5 - 2   =   12 - 5
=   3   =   7

Brackets are often used to show the order of a calculation. We go into this in more detail in Chapter 6.

Inverses

Finally, subtraction is the **inverse** of addition. This means that adding and subtracting the same number cancel each other out. For example, $4 + 3 - 3 = 4$. On a number line it looks like this:

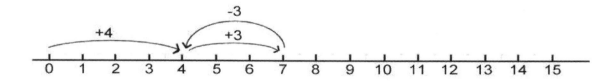

Adding and subtracting 0 makes no difference, so 0 is the **additive identity** number.

Adding and subtracting the same number gives the additive identity 0, so that + 3 and -3 are **additive inverses**.

Multiplication

Multiplication is repeated addition.

For example, here is three multiplied by four on a number line:

Three fours:

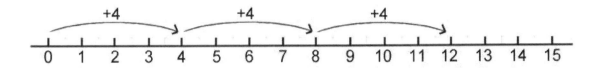

Or it could be four threes:

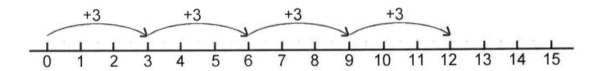

We use the words 'multiply' and 'times' to show multiplication and the symbol is 'x'. We also often imply multiplication such as 3 x 4 by simply saying 'three fours'. Also, the word 'of' implies multiplication in the sense that five crates of twelve bottles would make sixty bottles because 5 x 12 = 60.

The result of a multiplication is called a 'product'. For example, ten is the product of two and five. It is also a 'multiple' of two and a 'multiple' of five.

Like addition, it doesn't matter which in order you do several multiplications:

	From the left:			From the right:
	3 x 5 x 2 x 3			3 x 5 x 2 x 3
=	15 x 2 x 3		=	3 x 5 x 6
=	30 x 3		=	3 x 30
=	90		=	90

Also, like addition, you can reverse the order of a multiplication so that, for example, 4 x 5 = 5 x 4 = 20.

More formally, as multiplication is repeated addition, it inherits the associative and commutative properties so that, for example, (2 x 3) x 4 = 2 x (3 x 4) and 3 x 4 = 4 x 3.

However, the identity number is different. For addition it is 0, because 5 + 0 = 5. For multiplication, it is 1, because 5 x 1 = 5.

The **multiplicative identity** number is 1.

Squares

A particularly interesting type of multiplication occurs when you multiply a number by itself: 1 x 1, 2 x 2, 3 x 3 and so on. You get the sequence, 1, 4, 9, 16, and so on. These numbers are called the 'squares' because they give the area of squares with sides 1 unit long, 2 units long, and so on.

Squares occur often in physics and in engineering.

Squares are also an example of a type of number operation called 'raising to a power'. We will look at this later in Part 2.

Multiplication Tables

Strictly speaking, we need never do multiplication. We could always replace it with a repeated addition and this is actually what calculators and computers do. If you enter the calculation 439 x 354 into a calculator, it will add 439 over and over again 354 times. This may be inefficient, but a calculator does it so fast that you would never guess what it is doing.

This will not work for humans. It is too slow and we tend to make mistakes. We have to learn multiplication number bonds, usually called 'learning your tables'. Fortunately, we only have to learn multiplication facts up to 9 x 9. After that we use place value to break the calculation of larger multiplications into parts that can be added up to give the answer. In chapter 7, we will look at how this works and at several popular ways of laying out the calculation.

Here are some examples of multiplication tables:

Three Times				
1	x	3	=	3
2	x	3	=	6
3	x	3	=	9
4	x	3	=	12
5	X	3	=	15
6	X	3	=	18
7	X	3	=	21
8	X	3	=	24
9	X	3	=	27
10	X	3	=	30

Five Times				
1	x	5	=	5
2	x	5	=	10
3	x	5	=	15
4	x	5	=	20
5	X	5	=	25
6	X	5	=	30
7	X	5	=	35
8	X	5	=	40
9	X	5	=	45
10	X	5	=	50

Eight Times				
1	x	8	=	8
2	x	8	=	16
3	x	8	=	24
4	x	8	=	32
5	X	8	=	40
6	X	8	=	48
7	X	8	=	56
8	X	8	=	64
9	X	8	=	72
10	X	8	=	80

You can learn these by chanting them, if you like. When I was a child, my whole class was required to recite them together. It is a good start, but it doesn't go far enough. If you have to recite the table in order to recall what 7 x 8 is, you don't know it well enough.

The multiplication tables are often summarised in a 'table square', shown below.

X	0	1	2	3	4	5	6	7	8	9	10
0	0	0	0	0	0	0	0	0	0	0	0
1	0	1	2	3	4	5	6	7	8	9	10
2	0	2	4	6	8	10	12	14	16	18	20
3	0	3	6	**9**	**12**	**15**	**28**	**21**	**24**	**27**	30
4	0	4	8	12	**16**	**20**	**24**	**28**	**32**	**36**	40
5	0	5	10	15	20	**25**	**30**	**35**	**40**	**45**	50
6	0	6	12	18	24	30	**36**	**42**	**48**	**54**	60
7	0	7	14	21	28	35	42	**49**	**56**	**63**	70
8	0	8	16	24	32	40	48	56	**64**	**72**	80
9	0	9	18	27	36	45	54	63	72	**81**	90
10	0	10	20	30	40	50	60	70	80	90	100

(I will explain in the next page why some of these numbers are in bold.)

The table includes the multiples of 10 which are really not required because multiplying by ten is only a change of place value from units to tens, such as 3 x 10 = 30. I have also included the multiples of 0 and 1, simply for completeness. Many people learn their tables up to the twelve times, but now that we no longer have currency with twelve pennies to the shilling and no longer use dozens very much, there doesn't seem much point.

There is no easy way around the need to learn multiplication tables. You have got to bite the bullet and just do it. It is not as terrible as it might seem, because there is a lot of duplication. The bottom left half of the table is like a reflection of the top right half. For instance, if you know 3 x 4, then you can strip out 4 x 3. Once you have removed the mirror images such as 3 x 4 = 4 x 3 and all the very easy ones like the nought times, once times, two times tables and also the ten times table, there are only twenty eight facts to remember. I have shown these twenty eight facts in bold in the table square above, and here they are again in a reduced table square.

X	3	4	5	6	7	8	9
3	9	12	15	28	21	24	27
4		16	20	24	28	32	36
5			25	30	35	40	45
6				36	42	48	54
7					49	56	63
8						64	72
9							81

There are various ways to 'learn your tables'. I recommend flashcards. Get twenty eight small cards and write the multiplied numbers on one side and their answers on the other. For example, write 7 x 8 on one side and write 56 on the other side. Then shuffle the pack and read through them trying to remember what is on the other side of the card. If you can't remember, take a look before trying the next one. Keep on doing this until you have learnt them thoroughly. You might want someone else to help you with this. (numbersexplained.co.uk/downloads has printable flashcards.)

By thoroughly, I mean that as soon as you see 4 x 8, the number 32 immediately pops up in your mind. It has to be that automatic. You must also learn the tables backwards so that the moment you see 63 you immediately think 'nine sevens'.

If you can't do this, you will make mistakes in calculations involving multiplications and divisions of counting numbers. You may argue that you can always use a calculator, but that is like arguing that you don't need to be able to walk because you can always call a taxi. You will also find fractions very difficult and a calculator will not help.

Sorry, but there it is: learn them. You will be pleased when you have done it.

Numbers Explained

Division

Division is repeated subtraction. It is the answer to the question "How many times can I take a number away from another?"

The words that indicate division are 'divide', 'share' and 'by' and the sign is '÷'. However, because fractions involve dividing numbers into equal parts, the notation of fractions is often used to show division so that $10 ÷ 2$ is written as $\frac{10}{2}$ or 10/2.

The result of a division is called a 'dividend' or a 'quotient'. In a division, the number you are dividing by is called the 'divisor'.

Here is an example of a division, $15 ÷ 5 = 3$, shown on a number line. Start at 15 and see how many times you can subtract 5 before reaching 0 :

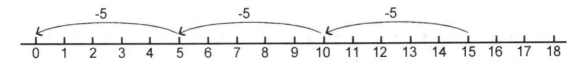

In the above example, we ended up at zero. In other words, 5 divides 15 exactly. Here is another example, $15 ÷ 6$, where it doesn't and there is a remainder:

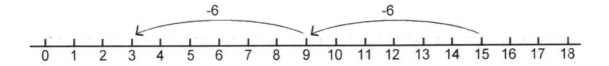

In the last example, we could take 6 away from fifteen two times but there was a remainder of 3 left over. The answer is often given as 2 remainder 3. This isn't a number. We can't deal easily with this problem if we use only counting numbers. The answer is to use fractions and this is the topic of later chapters.

Like subtraction, you cannot reverse the order of a division. $8 ÷ 2 ≠ 2 ÷ 8$. The first one, $8 ÷ 2$, gives 4 and the second one, $2 ÷ 8$, gives the fraction $\frac{1}{4}$. (If you are thinking that there must be a connection between 4 and $\frac{1}{4}$, you are right, but we will deal with that later.)

Also, you cannot do a series of divisions in any order. Like subtraction, you have to use brackets to show which comes first. Here is an example:

	$(16 \div 4) \div 2$			$16 \div (4 \div 2)$
=	$4 \div 2$		=	$16 \div 2$
=	2		=	8

In general, if you divide by anything other than a single number, you need brackets to determine the required order. For instance, $24 \div 4 \times 2$ looks innocent enough, but the order matters:

	$(24 \div 4) \times 2$			$24 \div (4 \times 2)$
=	6×2		=	$24 \div 8$
=	12		=	3

Finally, division is the inverse of multiplication. For example, $10 \times 2 \div 2 = 10$.

More formally, $\div 2$ is the **multiplicative inverse** of $\times 2$ because $2 \div 2 = 1$ which is the multiplicative identity number.

Dividing by Zero

You cannot divide by zero. If you try it on a calculator it will give an error message. The reason is clear if you remember that division is the answer to the question "How many times can I subtract this number?" You can carry on taking away 0 forever.

Programmers have to make sure that their instructions to the computer never cause an attempt to divide by zero, because this will cause a 'crash' as the computer uses up its memory trying to repeat a subtraction forever.

You cannot divide by zero.

How the 'Four Rules' are Connected

The diagram below gives a summary of the way the four rules of number are connected to each other.

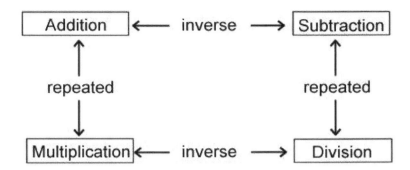

> **Addition is a movement up the number line and subtraction is an inverse movement the other way. Multiplication is repeated addition and division is the inverse – repeated subtraction.**

Odd and Even Numbers

The multiples of 2 (the numbers in the two times table, but going on to infinity) are called the set of even numbers. If a number isn't even, it is odd.

So all the counting numbers are alternately odd and even – odd, even, odd, even, odd and so on.

You can easily spot whether a number is odd or even. All odd numbers end with a 1, 3 5, 7 or 9 in the units place value, for example 17, 153, or 3 155 269. All even numbers end in 0, 2, 4, 6 or 8, for example 24, 182 or 173 440 376.

Knowing whether numbers are odd or even can be helpful in checking answers. The following rules apply to all counting numbers (but not to fractions or decimals)..

Addition and Subtraction

For addition and subtraction, you will always get an even answer if both numbers are even or odd. If one is odd and the other even, the answer will be odd.

Here are some examples:

4 + 6 = 10	(even + even = even)	12 − 4 = 8	(even − even = even)
3 + 12 = 15	(odd + even = odd)	27 − 12 = 15	(odd − even = odd)
8 + 7 = 15	(even + odd = odd)	32 − 3 = 29	(even − odd = odd)
11 + 7 = 18	(odd + odd = even)	19 − 9 = 10	(odd − odd = even)

Multiplication and Division

For multiplication and division, the answer will be even unless both numbers are odd.

Here are some examples:

4 x 6 = 24	(even x even = even)	32 ÷ 4 = 8	(even ÷ even = even)
7 x 6 = 42	(odd x even = even)	(you can't divide an odd by an even)	
8 x 7 = 56	(even x odd = even)	24 ÷ 3 = 8	(even ÷ odd = even)
9 x 7 = 63	(odd x odd = odd)	49 ÷ 7 = 7	(odd ÷ odd = odd)

All this might look hard to remember, but when you have practised arithmetic for a while, you do get a feel for it. If you see a calculation that violates these rules, it just looks wrong.

Summary of Odd and Even Rules

> **Addition (and subtraction): two evens or two odds makes an even; even and odd makes odd.**
> **Multiplication (and division): both odds make odd, otherwise even.**

Factors

A number that will divide exactly into another one is called a '**factor**'. For example, 2 is a factor of 8.

If 2 is a factor of 8, then so is 4 because 8 = 4 x 2. Factors always come in pairs like this.

Breaking a number down into factors is called factorisation.

Some numbers have lots of factors. For example, 60 is 1 x 60 , 2 x 30 , 3 x 20 , 4 x 15 , 5 x 12 and 6 x 10. So the factors of 60 are the numbers 1, 2, 3, 4, 5, 6, 10, 12, 15, 20, 30 and 60.

Factors are not just of theoretical interest. The ability to spot the factors of numbers is crucial when we start working with fractions. Sorry to keep harping on about this, but you do have to know your multiplication tables really well.

> **A counting number that will divide exactly into another one is called a 'factor'.**

It is possible to look at any division in two ways because its factors come in pairs. I said earlier that a division such as 12 ÷ 4 = 3 is asking the question "How many times can I take four away from twelve?" It can also be thought of as the question "What is the largest number that I can take away from twelve, four times?" This follows from the fact that multiplications can be reversed: 3 x 4 = 4 x 3. Here are both questions, shown on number lines:

How many times can I take 4 away from 12 ? Answer 3.

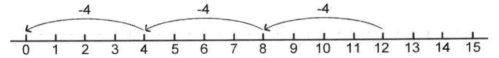

What is the largest number that I can take away from 12 four times? Answer 3.

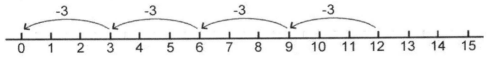

Prime Numbers

Some counting numbers have only one pair of factors. For example, 5 = 1 x 5 only.

Another way of putting this is that the only numbers that will divide them exactly are the number 1 and the number itself. Numbers like this are called **prime** numbers.

A prime number can only be divided by 1 and itself.

You would think that the first prime numbers would be 1, but 1 is excluded from the set of primes because of its special property: multiplying or dividing by 1 has no effect. So the first prime number is 2, which is the only even prime. All the rest are odd (because all even numbers must have a factor of 2, by definition). The remaining primes less than 100 are: 3, 5, 7, 11, 13, 17, 19, 23, 29, 31, 37, 41, 43, 47, 53, 59, 61, 67, 71, 73, 79, 83, 89, 97. Do you see a pattern? No? That's because, so far as anyone knows, there isn't one.

Prime numbers are of interest because they are the building blocks of all the other counting numbers, which are called **composite**. All composite numbers can be expressed as the unique product of two or more prime numbers.

For example, 60 = 2 x 2 x 3 x 5, 84 = 2 x 2 x 3 x 7, 935 = 5 x 11 x 17.

It can be proved that there are an infinite number of prime numbers (the proof is included in Part 2), but over the centuries, mathematicians have looked for and failed to find a practical way of deciding if a large number is prime. The problem is that dividing a number by all the numbers that might be factors takes far too long, even with modern computers. Some types of large numbers can be tested by other methods, but most large numbers remain undecided. Here is a problem which is readily understood, but for which there is no satisfactory answer. There are lots of problems like this in number theory.

A search for prime numbers is going on all the time. There is no space here to write the largest prime found so far (in January 2014) as it has 17,425,170 digits.

Chapter 6 – Mixing up the 'Four Rules' – BODMAS

In the last chapter we saw what happens when you have several numbers added or multiplied together. You can do the additions or multiplications in any order you like.

It is different for subtraction and division. The order matters.

For subtraction, $6 - 2 \neq 2 - 6$ and we must use brackets to distinguish between $(10 - 3) - 2$ and $10 - (3 - 2)$.

Similarly for division, $10 \div 5 \neq 5 \div 10$ and we must use brackets to distinguish between $(24 \div 4) \div 2$ and $24 \div (4 \div 2)$.

This chapter looks at the rules that apply when you mix up the four rules of number. For example, how do you do 10 + 3 x 4 and is it the same as 3 x 4 + 10?

First, some vocabulary:

Arithmetic Expressions

The signs +, -, x, and ÷ are called arithmetic **operators**. They tell us what to do. There are some other operators that we will deal with later, but for now we will just use operators for the 'four rules' and use brackets like these **( )** to give us any ordering that is needed.

A meaningful combination of numbers and operators is called an arithmetic **expression**. Here are some examples:

    2 + 3

    12 x 3 – 5

    40 ÷ 5 + (4 + 6 x 2) – 4 x 7

This chapter is all about working out the answers to arithmetic expressions like these

Do Not Work From Left to Right

Arithmetic expressions are not always worked out from left to right.

This may come as a bit of a surprise, but here is a proof that working from left to right doesn't always work.

Working from left to right, like this:

$$3 + 5 \times 2$$
$$= \quad 8 \times 2$$
$$= \quad 16$$

But we know that multiplication is repeated addition, so we could do it by replacing 5 x 2 with 5 + 5, like this:

$$3 + 5 \times 2$$
$$= \quad 3 + 5 + 5$$
$$= \quad 13$$

Well, which is it? It can't be both.

The difficulty is sorted out if we stop reading from left to right and do the multiplication first, like this:

$$3 + 5 \times 2$$
$$= \quad 3 + 10$$
$$= \quad 13$$

The answer is, indeed, 13.

There are some rules that govern the mixing up of operators and we will start by looking at just addition and multiplication.

Mixing Addition and Multiplication

If you have an arithmetic expression that mixes up addition and multiplication, you must do the multiplication first, or you will get the wrong answer. The reason is that multiplication is a whole lot of additions bundled up together. It is possible to 'unbundle' the multiplication and do the whole calculation by addition (like a calculator but slower, so not recommended). Otherwise, you must do the multiplication first.

Here is another example done firstly by 'unbundling' the multiplication and then by following the 'multiplication first' rule:

	5 + 6 x 4				5 + 6 x 4
=	5 + 6 + 6 + 6 + 6		=		5 + 24
=	29		=		29

From now on we will leave the 'unbundling' approach and use 'multiplication first'.

What do arithmetic expressions like these mean in the real world? Multiplication is often used to count things that are laid out in rectangular patterns of rows and columns. Any rectangular pattern will do, but I like to use diagrams of cabbage patches in a vegetable garden.

Here is 3 + 5 x 2:

Think of this as 'three cabbages and two rows of five cabbages'.

Here is 5 + 6 x 4:

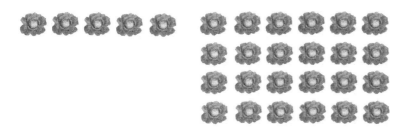

Think of this as 'five cabbages and four rows of six cabbages'.

But suppose that we put in some brackets because we actually want to do the addition first. What does that mean?

Here are the two previous expressions changed into (3 + 5) x 2 and (5 + 6) x 4. They are shown worked out and with cabbage patch illustrations.

```
      (3 + 5) x 2
  =   8 x 2
  =   16
```

Think of this as 'two rows of three and five cabbages'.

```
      (5 + 6) x 4
  =   11 x 4
  =   44
```

Think of this as 'four rows of five and six cabbages'.

All this can be summarised into a rule that you **must** follow when working out arithmetic expressions:

> **Do anything in brackets first, then any multiplications, and finally any additions.**

Here are two similar arithmetic expressions, with and without brackets around the addition:

```
      5 + 3 x 4              (5 + 3) x 4
  =   5 + 12             =   8 x 4
  =   17                 =   32
```

Here are four longer arithmetic expressions that give different answers depending upon where any brackets are placed. I have put a brief explanation at the end of each line. Note that you can do more than one operation at the same time, so long as you stay with the order: 1) brackets, 2) multiply, 3) add. Please examine them carefully.

2 + 3 x 4 + 5 (no brackets so do the multiplication)

=   2 + 12 + 5 (now do the additions)

=   19

(2 + 3) x 4 + 5 (do the bracket)

=   5 x 4 + 5 (now do the multiplication)

=   20 + 5 (now do the addition)

=   25

2 + 3 x (4 + 5) (do the bracket)

=   2 + 3 x 9 (now do the multiplication)

=   2 + 27 (now do the addition)

=   29

(2 + 3) x (4 + 5) (do both brackets)

=   5 x 9 (now do the multiplication)

=   45

Please note the way that these calculations are laid out (ignoring the explanation comments). Each line has the full expression written down after an equals sign (so that it really is equal) but with part of the calculation worked out. I strongly recommend that you follow this layout, for two reasons. Firstly, it is very logical and stops you from getting confused about what to do next. Secondly, it will be necessary if you decide to do any algebra later and you may as well acquire some good habits now.

You can have a mixture of multiplication and addition inside each bracket. Consider these four different placings of brackets:

2 x 4 + 3 x 5 (no brackets, do the multiplications)

=   8 + 15 (now do the addition)

=   23

2 x (3 + 4) x 5 (do the bracket)

=   2 x 7 x 5 (now do the multiplications)

=   70

(2 x 4 + 3) x 5 (multiplication inside the bracket)

=   (8 + 3) x 5 (addition inside the bracket)

=   11 x 5 (now do the multiplication)

=   55

2 x (4 + 3 x 5) (multiplication inside the bracket)

=   2 x (4 + 15) (addition inside the bracket)

=   2 x 19 (now do the multiplication)

=   38

Arithmetic expressions like these can be as long and complicated as you like. Just follow the rule: 1) brackets, 2) multiply, 3) add.

Now we will fit subtraction and division into these rules.

Subtraction and Division

Subtraction is the inverse of addition. Whereas addition is a movement to the right on a number line, subtraction is a movement to the left. So we can do additions and subtractions at the same time. The only thing to watch out for are repeated subtractions like $10 - 3 - 2$. This needs an extra pair of brackets to show whether we mean:

	$(10 - 3) - 2$	or		$10 - (3 - 2)$
=	$7 - 2$		=	$10 - 1$
=	5		=	9

Otherwise, it is quite straightforward: do additions and subtractions together.

This means that our rule becomes: 1) brackets, 2) multiply, 3) add and subtract.

The rule for division is similar. Division is the inverse of multiplication, so we can do it at the same time as multiplication, provided we look out for repeated divisions like $24 \div 4 \div 2$ which could mean $(24 \div 4) \div 2 = 3$ or $24 \div (4 \div 2) = 12$.

So, the rule now becomes: 1) brackets, 2) multiply and divide, 3) add and subtract.

Here are some examples like the previous ones but including subtractions and divisions. However, they have been carefully chosen to give counting number answers. If you try to create some examples of your own (and I would encourage you to do so), you may well end up with negative or fraction answers. These are topics that we have mentioned but not yet looked at thoroughly.

	$8 + 12 \div 4 - 2$ (no brackets so do the division)			$(8 + 12) \div 4 - 2$ (do the bracket)
=	$8 + 3 - 2$ (now do the addition and subtraction)		=	$20 \div 4 - 2$ (now do the division)
=	9		=	$5 - 2$ (now do the subtraction)
			=	3

	$8 + 12 \div (4 - 2)$ (do the bracket)			$(8 + 12) \div (4 - 2)$ (do both brackets)
=	$8 + 12 \div 2$ (now do the division)		=	$20 \div 2$ (now do the division)
=	$8 + 6$ (now do the addition)		=	10
=	14			

Numbers Explained

Here are some more complicated examples including mixed operations inside the brackets:

	2 x 12 - 8 ÷ 2 (both multiplication and division)			2 x (12 - 8) ÷ 2 (do the bracket)
=	24 - 4 (now do the subtraction)		=	2 x 4 ÷ 2 (now do the multiplication and division)
=	20		=	4

	(2 x 12 - 8) ÷ 2 (multiplication inside the bracket)			2 x (12 - 8 ÷ 2) (division inside the bracket)
=	(24 - 8) ÷ 2 (subtraction inside the bracket)		=	2 x (12 - 4) (subtraction inside the bracket)
=	16 ÷ 2 (now do the division)		=	2 x 8 (now do the multiplication)
=	8		=	16

As you can see, if you take the trouble to write down each step in full and follow the rule, it isn't that hard. If you try to do anything other than a very simple example in your head, you will probably get it wrong.

BODMAS

The rule that we have seen above is usually remembered by the rather strange word 'bodmas'. (Memory words like this are called mnemonics.) The letters stand for:

Brackets **O**f **D**ivide **M**ultiply **A**dd **S**ubtract

Where did 'Of' come from. It is another word for multiplication and is only there to make the word more pronounceable. Otherwise it would be 'bdmas'. (You will also see 'bidmas' where the 'i' stands for 'indices' which will be covered in Part 2.) Really, there are three steps: and so it is best to group the letters like this:

1) **B**rackets 2) **O**f **D**ivide **M**ultiply 3) **A**dd **S**ubtract

If it helps, remember 'BODMAS'.

In 1924, Jan Łukasiewicz invented a way of writing arithmetic expressions called Polish notation that does away with brackets. The operators go <u>before</u> the numbers so that 2 + 3 is written + 2 3. With more complex expressions, the rule is that an operator is only used if it has two numbers immediately following it. This means that + x 3 4 5 = + 12 5 = 17, whereas + 3 x 4 5 = + 3 20 = 23. It seems odd at first, but it works very well. It hasn't caught on except in computer programming.

Expanding Brackets

There is another way of working out answers with brackets. This will lead us to a way of multiplying large numbers.

Suppose that we have a cabbage patch that looks like this:

You can think of this in two ways:

Two rows of three and five cabbages: (3 + 5) x 2

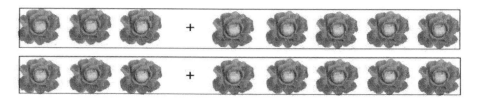

Or   Two rows of three and two rows of five: 3 x 2 + 5 x 2

In the first expression, everything being added inside the brackets is being multiplied by 2. So, this is the same as two separate multiplications added together.

This means that there are two ways of getting the answer:

	(3 + 5) x 2	gives the same answer as		(3 + 5) x 2
=	3 x 2 + 5 x 2		=	8 x 2
=	6 + 10		=	16
=	16			

Or, with the multiplication the other way round:

	2 x (3 + 5)	gives the same answer as		2 x (3 + 5)
=	2 x 3 + 2 x 5		=	2 x 8
=	6 + 10		=	16
=	16			

This procedure is called **expanding** the brackets. Everything inside the bracket gets multiplied.

It works for more than two numbers inside the brackets:

	(3 + 5 + 4) x 2	gives the same answer as		(3 + 5 + 4) x 2
=	3 x 2 + 5 x 2 + 4 x 2		=	12 x 2
=	6 + 10 + 8		=	24
=	24			

It also works for subtraction:

	(12 - 5) x 2	gives the same answer as		(12 - 5) x 2
=	12 x 2 - 5 x 2		=	7 x 2
=	24 - 10		=	14
=	14			

And it also works for division one way:

	(4 + 8) ÷ 2	gives the same answer as		(4 + 8) ÷ 2
=	4 ÷ 2 + 8 ÷ 2		=	12 ÷ 2
=	2 + 4		=	6
=	6			

But note that it **doesn't** work for division the other way round:

	12 ÷ (4 + 2)	can't be expanded this way and		12 ÷ (4 + 2)
≠	12 ÷ 4 + 12 ÷ 2 **wrong!**	does not give the same answer as	=	12 ÷ 6
=	3 + 6		=	2
=	9			

Now comes the interesting bit. You can also do an expansion of brackets backwards. For example, on the next page:

	8 x 6	or, breaking the 6		8 x 6
=	(5 + 3) x 6		=	8 x (4 + 2)
=	5 x 6 + 3 x 6		=	8 x 4 + 8 x 2
=	30 + 18		=	32 + 16
=	48		=	48

Why would you want to do that? Here's why.

In the following example, one of the numbers in a multiplication is going to be broken down into the sum of two numbers inside a bracket:

	5 x 15
=	5 x (10 + 5)
=	5 x 10 + 5 x 5
=	50 + 25
=	75

You may or may not already know that 5 x 15 is 75, but notice what we have done. We have broken 5 x 15 into two very easy multiplications: 5 x 10 and 5 x 5. You can easily work these out and then add the answers together.

This way of breaking down a multiplication is the basis of the multiplication methods for large numbers. It depends upon breaking the calculation down into parts by place value. (In the simple example above we broke 15 down into tens and units.) Very similar methods are also used with the division of large numbers. The next chapter looks at these techniques.

Negative Signs Before Brackets

Brackets can also be used to show repeated subtractions. Here is an example:

$$20 - 4 - 3$$

This is an ambiguous expression. It could mean either

	(20 − 4) − 3			20 − (4 − 3)
=	16 − 3	or	=	20 − 1
=	13		=	19

Numbers Explained

Usually, we intend the first meaning, "subtract 4 and then 3 from 20". The second meaning, "subtract the difference between 4 and 3 from 20" is less likely.

Another way of expressing the first meaning is to use brackets to enclose the sum of the numbers to be subtracted, like this:

$$20 - (4 + 3)$$
$$= \quad 20 - 7$$
$$= \quad 13$$

The minus sign in front of the bracket means that <u>everything</u> inside the bracket is to be subtracted.

Brackets Inside Brackets

It is possible to have complicated expressions that have to have two sets of brackets, one nested inside the other. It is a good idea to use different types of bracket so that you can see which opening and closing brackets are paired with which.

You have to work out the inside bracket first.

Here is an example:

$$50 - [(100 \div 2) \div 5 - 4] \quad \text{(do the inside bracketed division: } 100 \div 2)$$
$$= \quad 50 - [50 \div 5 - 4] \quad \text{(now do the other inside division: } 50 \div 5)$$
$$= \quad 50 - [10 - 4] \quad \text{(now do the subtraction in the outer bracket)}$$
$$= \quad 50 - 6 \quad \text{(finally do the outside subtraction)}$$
$$= \quad 44$$

This sort of calculation is not common in everyday arithmetic, but you will encounter nested sets of brackets if you go on to do any algebra.

Practice Exercises

You can download practice examples of mixed operations from http://numbersexplained.co.uk.

Chapter 7 – The 'Four Rules' for Larger Numbers

So far we have kept the numbers small when dealing with the 'four rules'. This chapter explains how to use place value to perform calculations with large numbers.

If you are happy to rely on a calculator, you can skip this chapter.

Addition

Addition is done by carefully organising the numbers into columns. If you want to calculate 34 + 21 + 30, write it down as:

```
    3   4
    2   1
    3   0
  _____
```

Then add up each column, 4 + 1 + 0 = 5 and 3 + 2 + 3 = 8:

```
    3   4
    2   1
    3   0
    8   5
  _____
```

So the answer is 85.

What happens if the sum of each column comes to more than 9? Suppose we have 54 + 78:

```
    5   4
    7   8
  _____
```

It is important to start with the units in the right-hand column: 4 + 8 = 12. The digit 2 will go in the units place value of the answer, but the digit 1 has a place value of ten and must go in its correct column. So we 'carry' the 1 into the tens column.

The usual place to write this is under the bottom line, like this:

```
    5   4
    7   8
  _____
        2
  _____
    1
```

Now we can add up the tens, not forgetting the 'carried' 1, 5 + 7 + 1 = 13. The digit 3 will go in the tens place value and we will start another column for hundreds to 'carry' the digit 1. Since there are no hundreds in the original sum, we don't have to write this 'carry' under the bottom line. We can just write it in the answer:

```
      5   4
      7   8
    _____
  1   3   2
    _____
      1
```

The answer is 132.

That's it. Very large numbers just have more columns. You must be careful to line them up properly and make sure that you don't miss out any zeros. For example: 135 450 + 29 031 is laid out as:

```
  1   3   5   4   5   0
      2   9   0   3   1
  _____
  1   6   4   4   8   1
  _____
      1
```

The answer is 161 481.

Addition only becomes tricky if you have a lot of numbers to add up. Firstly, you have to be able to add up each column in your head and this is a skill that takes practice.

Secondly, it is possible that you may have to 'carry' into more than one column. Here is an example:

```
      8  9
      8  8
   1  9  9
      4  8
   1  0  9
      9  7
      9  8
      8  8
      9  7
      9  9
   1  0  3
      8  9
      9  8
   1  7  8
   _____
            0
   _____
   1  1
```

The units column adds up to 110. This means that in the answer there will be a 0 in the units column and there will be a 1 to carry into the tens and another 1 to carry into the hundreds.

```
      8  9
      8  8
   1  9  9
      4  8
   1  0  9
      9  7
      9  8
      8  8
      9  7
      9  9
   1  0  3
      8  9
      9  8
   1  7  8
   _____
      7  0
   _____
   1  1
   9
```

Now the tens column adds up to 97 plus the carried 1 makes 98. So an 8 goes in the tens column and a 9 is carried in to the hundreds. The best place to write this is under the existing carried 1.

```
                8   9
                8   8
            1   9   9
                4   8
            1   0   9
                9   7
                9   8
                8   8
                9   7
                9   9
            1   0   3
                8   9
                9   8
            1   7   8
    _____
    1   4   8   0
    _____
            1   1
            9
```

Finally add up the hundreds, which is 1 + 1 + 1 + 1 = 4 plus the carried 1 + 9 makes 14. The 4 goes in the hundreds. Then start a thousands column, and carry the 1 into it.

The answer is 1 480.

Adding a lot of numbers together like this is not easy and requires careful checking. You should only attempt it if you don't have a calculator handy.

A good way of checking each column is to add it up twice, once from top to bottom and then from bottom to top and see if you get the same total.

Subtraction

Like addition, subtraction must be laid out with the place values in neat columns. Then subtract each column. For example, 87 - 35:

```
    8   7
    3   5
  _____
    5   2
  _____
```

The answer is 52.

The only tricky bit is when the subtraction in a column gives you a negative answer. Here is an example: $63 - 27$:

6	3
2	7

63 is larger than 27, so there is an answer that will be a positive counting number. However, we have $3 - 7$ in the units place value, which will be negative. Now what?

There are two ways of dealing with this. The method that is favoured now is called 'decomposition' but older people may use a method called 'equal additions'. Both methods will be explained now. They are equally good and you can decide which one you would like to use.

It is important in both methods that you start with the units column and work through the columns from right to left.

Subtraction by Decomposition

This depends upon changing 63 into $50 + 13$. The usual way to lay it out is like this:

6	3
2	7

Cross out the 6 and write 5, and put a little 1 next to the 3 to make it 13, so it becomes:

6^5	^{13}
2	7

Now we have $13 - 7$ instead of $3 - 7$ in the units place value and the tens place value has become $5 - 2$ instead of $6 - 2$. Both of these will subtract without giving a negative answer.

6^5	^{13}
2	7
3	6

The answer is 36.

It may be necessary to use this decomposition trick more than once. For example:

3	5	2
	8	9

becomes

3	5^4	^{12}
	8	9
		3

The units place value is now 12 - 9, giving an answer of 3. But now the trick must be repeated in the tens place value because 4 − 8 is also negative.

$$3 \quad \cancel{5}^{4} \; {}^{1}2$$
$$\underline{\qquad 8 \quad 9} \qquad \text{becomes}$$

$$\cancel{3}^{2} \; \cancel{5}^{14} \; {}^{1}2$$
$$\underline{\qquad\quad 8 \quad 9}$$
$$\qquad\qquad\quad 3$$

Now 14 − 8 in the tens column gives 6 and then 2 − 0 in the hundreds column gives 2 to complete the calculation.

$$\cancel{3}^{2} \; \cancel{5}^{14} \; {}^{1}2$$
$$\underline{\qquad\quad 8 \quad 9}$$
$$\underline{\quad 2 \quad 6 \quad 3}$$

The answer is 263.

This decomposition method has the merit of being easy to understand. The only problem is that it can become complicated when you have several columns and the values in each column are unhelpful. Here is an example.

$$1 \quad 3 \quad 0 \quad 3$$
$$\underline{\quad\; 5 \quad 4 \quad 7}$$
$$\underline{\qquad\qquad\qquad}$$

In the units, 3 − 7 is negative so we need one of the tens to give us 13, but there are no tens to decompose. The solution is to decompose one of the hundreds:

$$1 \quad 3 \quad 0 \quad 3$$
$$\underline{\quad\; 5 \quad 4 \quad 7} \qquad \text{becomes}$$
$$\underline{\qquad\qquad\qquad}$$

$$1 \quad \cancel{3}^{2} \; {}^{1}0 \quad 3$$
$$\underline{\quad\quad\; 5 \quad 4 \quad 7}$$
$$\underline{\qquad\qquad\qquad}$$

Now there are 10 tens. This is now decomposed into 9 tens and 10 in the units to give us the 13 we need:

$$1 \quad \cancel{3}^{2} \; \cancel{{}^{1}0}^{9} \; {}^{1}3$$
$$\underline{\quad\;\; 5 \qquad 4 \quad 7} \qquad \text{do } 13 - 7 \text{ in the units and}$$
$$\underline{\qquad\qquad\qquad\qquad} \qquad \text{also } 9 - 4 \text{ in the tens to get}$$

$$1 \quad \cancel{3}^{2} \; \cancel{{}^{1}0}^{9} \; {}^{1}3$$
$$\underline{\quad\;\; 5 \qquad 4 \quad 7}$$
$$\underline{\qquad\qquad\; 5 \quad 6}$$

But now we have $2 - 5$ in the hundreds place value, so the decomposition trick is required again:

$$1 \quad 3^2 \quad {}^1\!\cancel{0}^9 \quad {}^1 3$$
$$\qquad\quad 5 \quad 4 \quad 7 \qquad \text{becomes}$$
$$\overline{\qquad\qquad\quad 5 \quad 6}$$

$$\cancel{1}^0 \quad 3^{12} \quad {}^1\!\cancel{0}^9 \quad {}^1 3$$
$$\qquad\qquad 5 \quad 4 \quad 7$$
$$\overline{\qquad\qquad\qquad 5 \quad 6}$$

Now we have $12 - 5$ in the hundreds and nothing left at all in the thousands so the completed calculation looks like this:

$$\cancel{1}^0 \quad 3^{12} \quad {}^1\!\cancel{0}^9 \quad {}^1 3$$
$$\qquad\qquad 5 \quad 4 \quad 7$$
$$\overline{\qquad\quad 7 \quad 5 \quad 6}$$

The answer is 756.

It doesn't get much more complicated than this, but I hope that you can see the need to be very neat in the layout or else it is very easy to make mistakes.

Subtraction by Equal Additions
Equal additions, as the name suggests, works by adding the same number to both numbers in the subtraction. Using our first example again, $63 - 27$, it works by changing 63 into 73 in the form of the addition $60 + 13$ and changing 27 into 37. This gives the right answer because the answers to $63 - 27$ and $73 - 37$ are the same.

Here's how it is laid out:

$$6 \quad 3$$
$$2 \quad 7 \qquad \text{becomes}$$
$$\overline{\qquad\qquad}$$

$$6 \quad {}^1 3$$
$${}^3 2 \quad 7$$
$$\overline{\qquad\qquad}$$

In effect, 10 has been added to both numbers, in the units column in the top line and in the tens column in the bottom line. Now the columns are subtracted to give:

$$6 \quad {}^1 3$$
$${}^3 2 \quad 7$$
$$\overline{3 \quad 6}$$

The answer, as before, is 36.

Here is the second example, 352 -89, where the trick has to be used twice:

$$\begin{array}{ccc} 3 & 5 & 2 \\ & 8 & 9 \\ \hline \\ \hline \end{array} \quad \text{becomes} \quad \begin{array}{ccc} 3 & 5 & {}^{1}2 \\ & {}^{9}8 & 9 \\ \hline & & 3 \\ \hline \end{array}$$

Now we have 5 – 9 in the tens place value, so we add 10 tens to the top number and 1 hundred to the bottom one:

$$\begin{array}{ccc} 3 & {}^{1}5 & {}^{1}2 \\ 1 & {}^{9}8 & 9 \\ \hline 2 & 6 & 3 \end{array}$$

The answer, as before, is 263.

Some people find this easier than decomposition because you can always add a 1 to the column on the left and so you never have to go the columns further left to make it work. However, it can get a bit messy to lay out unless you allow plenty of room between the columns, as in the following example shown fully worked out:

$$\begin{array}{ccc} 2 & {}^{1}2 & {}^{1}7 \\ {}^{2}1 & {}^{10}9 & 8 \\ \hline & 2 & 9 \end{array}$$

The answer, after much crossing out and writing of tiny figures, is 29.

The last example offers a chance to use a method of subtraction that you can do in your head called 'counting on'. Here's how it works:

Start with 198, add 2 to make 200 and then another 27 makes 227. We have added 2 and 27, which adds up to 29.

In this example, this is much easier than either of the written methods of subtraction. You could also use counting on to calculate any subtraction provided that you devised a way of writing the procedure down in some sort of table.

Multiplication

If you have a multiplication that is too large to do in your head, you have to break it down into separate multiplications for each place value and then add up the results to get the final answer. We will tackle this in two stages, firstly, when just one of the numbers is too big to handle on its own, and secondly, when both numbers are too large.

Firstly, you need to be clear that if, for instance, 3 x 4 = 12, then 3 x 40 = 120 and 3 x 400 = 1200, and so on. It is the same multiplication but in different place values. We will now use this when multiplying large numbers.

In principle, here is what to do when one number is too large, for example, 4 x 357. I don't know the four times table up to 357 times and I don't suppose that you do either. However, we can break up 357 into 300 + 50 + 7 so that we get three separate multiplications, like this:

    4 x 357
=   4 x (300 + 50 + 7)
=   4 x 300 + 4 x 50 + 4 x 7

These three separate multiplications are possible. We know that 4 x 3 = 12 and so 4 x 300 = 1200. We also know that 4 x 5 = 20 and so 4 x 50 = 200. Finally, 4 x 7 = 28. So, continuing the calculation:

    4 x 300 + 4 x 50 + 4 x 7
=   1200 + 200 + 28
=   1428

So, the answer to 4 x 357 is 1428.

We need a better way of laying this out and we will consider various methods when we have looked at what happens when both numbers are too large. What follows shows <u>in principle</u> what is going on when you multiply two large numbers – it isn't an efficient method and I am not suggesting that you should try it.

Suppose we want to know, not 4 x 357, but 629 x 357. This is broken up into nine separate multiplications, like this:

629 x 357

= 629 x (300 + 50 + 7)   (break up the 357)

= 629 x 300 + 629 x 50 + 629 x 7 (three products)

= (600 + 20 + 9) x 300
 + (600 + 20 + 9) x 50   (now break up the 629)
 + (600 + 20 + 9) x 7

= 600 x 300 + 20 x 300 + 9 x 300
 + 600 x 50 + 20 x 50 + 9 x 50   (nine products)
 + 600 x 7 + 20 x 7 + 9 x 7

= 180 000 + 6 000 + 2 700
 + 30 000 + 1 000 + 450   (now add them up)
 + 4 200 + 140 + 63

= 224 553

Working for the addition:

$$
\begin{array}{r r}
1\ 8\ 0 & 0\ 0\ 0 \\
6 & 0\ 0\ 0 \\
2 & 7\ 0\ 0 \\
3\ 0 & 0\ 0\ 0 \\
1 & 0\ 0\ 0 \\
 & 4\ 5\ 0 \\
4 & 2\ 0\ 0 \\
 & 1\ 4\ 0 \\
 & 6\ 3 \\
\hline
2\ 2\ 4 & 5\ 5\ 3 \\
\end{array}
$$

There are various ways of laying this out that take less space. Some of them also automatically take care of the place value in each separate multiplication. However, they are all based on the principle shown above.

Table Layout

This is popular in primary schools because it is fairly close to the explanation that we have just seen. Here is 629 x 357 again:

	600	20	9	
300	180 000	6 000	2 700	188 700
50	30 000	1 000	450	31 450
7	4 200	140	63	4 403
				224 553

There are variations on this. You can, of course, have 629 as the rows and 357 as the columns. You can add up the columns first and then add up their totals to get the answer. You may well need to write down some separate working to do this.

Lots of people find the table layout a perfectly good method and stay with it. Its advantage is that it is easy to understand. You can get the correct place value for each separate multiplication by 'adding the noughts'. The drawback is that it gives little help with adding across the rows. You can see in this example that the middle row is easy because there is no 'carrying'. The top row has one 'carry' to do, and the bottom row is tricky, with two 'carries' to do in your head.

Lattice Layout

This is a version of the table layout. It handles the place value automatically.

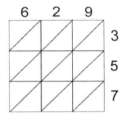

You draw a table that is large enough for all the separate multiplication, just like the one above. Ignoring the place value, head the columns 6, 2 and 9 and head the rows 3, 5 and 7 on the right of each row. Draw diagonal lines across the table.

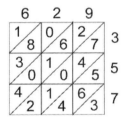

Now write the answer to each multiplication in the table so that the units part is below the diagonal and the tens part is above. If there are no tens, put in a zero.

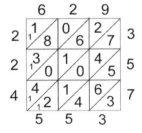

Finally, add down the diagonals and write the answers below and to the left of the table. Start with the bottom right corner of the table, and if there are any tens in each sum, carry it into the next diagonal and write it in a convenient space – you will probably have to write it small, as shown here. This 'carry' must be included in the sum of the diagonal.

Then read the answer 224 553 from the left and bottom edges of the table

The lattice method is very old. Originating in Hindu-Arabic mathematics, it was introduced to Europe by the Italian mathematician Fibonacci in a book published in 1202. It was later popularised by the 17th century Scottish mathematician John Napier, famous as the inventor of logarithms (explained in Part 2). He invented a type of calculator based on the lattice method using strips of wood called Napier's Bones.

The lattice method is a good foolproof way of multiplying numbers. Its only drawback is the need to draw a lattice that may be quite large.

The Standard Method

This layout is the most compact but requires some care in writing numbers in their correct place values.

Using our example, 629 x 357, here's how it works, step by step with each step shown in bold:

```
  6  2  9
  3  5  7
 _____

 _____
 _____
```

Write the numbers one above the other. Draw a line underneath, leave space for three lines of working. These three lines of working will contain the values of 7 x 629, 50 x 629 and 300 x 629. Draw another longer line. The answer will go under this line. Draw another line under that so that 'carry' digits can be written underneath.

```
  6  2  9
  3  5  7
 _____
        6  3
 _____
 _____
```

Multiply 7 x 9 which is 63, and write the 3 in the right-hand column of the first line of working. At the moment, the 6 of 63 can't go straight into the tens column to the left, because the result of the next multiplication has got to go there as well. Write it as a small 'carry digit' in this column.

```
  6  2  9
  3  5  7
 _____
     2  0⁶ 3
 _____
 _____
```

Multiply 7 x 2. This 2 is, of course, actually 20, but you don't have to think about its true place value so long as you put it into the next column. 7 x 2 = 14, but you must remember to add the carried 6, so that makes 20. Write in 0 and carry the 2 into the next column

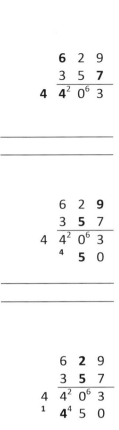

$$\begin{array}{cccccc} & & 6 & 2 & 9 \\ & & 3 & 5 & \mathbf{7} \\ 4 & \mathbf{4}^2 & 0^6 & 3 \end{array}$$

Multiply 7 x 6 which is 42, but add the carried 2 to make 44. Write this in the next two columns. There is no more working on this line, so the 'carried' 4 can be written down large.

What we have here is the result of 7 x 629.

$$\begin{array}{ccccc} & & 6 & 2 & \mathbf{9} \\ & & 3 & \mathbf{5} & 7 \\ 4 & 4^2 & 0^6 & 3 \\ & ^4 & \mathbf{5} & 0 \end{array}$$

The second line of working will be 5 x 629. You work your way along the top line from right to left, adding in any 'carried' digits as you go. The crucial thing is to start in the correct column. You are multiplying by 5, but it is actually 50, so you must start in the tens column and put 0 in the empty units.

First step: 5 x 9 = 45, write 5 and carry 4.

$$\begin{array}{ccccc} & & 6 & \mathbf{2} & 9 \\ & & 3 & \mathbf{5} & 7 \\ 4 & 4^2 & 0^6 & 3 \\ ^1 & \mathbf{4}^4 & 5 & 0 \end{array}$$

Second step: 5 x 2 = 10 plus the carried 4 makes 14, write 4 and carry the 1.

$$\begin{array}{ccccc} & 6 & 2 & 9 \\ & \mathbf{3} & \mathbf{5} & 7 \\ 4 & 4^2 & 0^6 & 3 \\ 3 & 1^1 & 4^4 & 5 & 0 \end{array}$$

Third step: 5 x 6 = 30 plus the carried 1 makes 31. Write this in.

That completes 5 times 629 (actually 50 x 629).

$$\begin{array}{cccccc} & & 6 & 2 & 9 \\ & & \mathbf{3} & 5 & 7 \\ & 4 & 4^2 & 0^6 & 3 \\ 3 & 1^1 & 4^4 & 5 & 0 \\ 1 & 8 & 8^2 & 7 & 0 & 0 \end{array}$$

Now the third line of working is 3 x 629, but because it is actually 300, not 3, you must start in the hundreds column and put zeros in the tens and units. Work along the top line, just as before.

Here is the whole line of working.

$$\begin{array}{ccccccc}
 & & & 6 & 2 & 9 \\
 & & & 3 & 5 & 7 \\
\hline
 & & 4 & 4^2 & 0^6 & 3 \\
 & 3 & 1^1 & 4^4 & 5 & 0 \\
1 & 8 & 8^2 & 7 & 0 & 0 \\
\hline
2 & 2 & 4 & 5 & 5 & 3 \\
 & & 1 & 1
\end{array}$$

Finally, add up the three lines of working as shown here. Do not add in the carried digits in the lines of working, because they have already been added in.

There are several variations on this above method. Here are three:

$$\begin{array}{ccccccc}
 & & & 6 & 2 & 9 \\
 & & & 3 & 5 & 7 \\
\hline
 & & 4 & 2^1 & 4^6 & 3 \\
 & 3 & 0^1 & 0^4 & 5 & 0 \\
1 & 8 & 6^2 & 7 & 0 & 0 \\
\hline
2 & 4 & 4 & 5 & 5 & 3 \\
 & & 1 & 1 & 1
\end{array}$$

Don't add in the carried digits in each line of working, but add them in when you do the final addition.

$$\begin{array}{ccccccc}
 & & & 6 & 2 & 9 \\
 & & & 3 & 5 & 7 \\
\hline
1 & 8 & 8^2 & 7 & 0 & 0 \\
 & 3 & 1^1 & 4^4 & 5 & 0 \\
 & & 4 & 4^2 & 0^6 & 3 \\
\hline
2 & 2 & 4 & 5 & 5 & 3 \\
 & & 1 & 1
\end{array}$$

Do the three lines of working in the reverse order.

$$\begin{array}{ccccccc}
 & & & 6 & 2 & 9 \\
 & & & 3 & 5 & 7 \\
\hline
 & & 4 & 4 & 0 & 3 \\
 & 3 & 1 & 4 & 5 & 0 \\
1 & 8 & 8 & 7 & 0 & 0 \\
\hline
2 & 2 & 4 & 5 & 5 & 3 \\
 & & 1 & 1
\end{array}$$

If you are good enough with your tables, you can do the carrying in your head as you go along each line of working.

You mustn't be interrupted as you do this, or you will have to start again.

Finally, of course, you could do the whole multiplication the other way round with 357 above 629 using any of these variations. Here is one of them:

$$
\begin{array}{rrrrrr}
 & & 3 & 5 & 7 & \\
 & & 6 & 2 & 9 & \\
\hline
 & 3 & 2^5 & 1^6 & 3 & \\
 & 7^1 & 1^1 & 4 & 0 & \\
2 & 1^3 & 4^4 & 2 & 0 & 0 \\
\hline
2 & 2 & 4 & 5 & 5 & 3 \\
 & 1 & & & &
\end{array}
$$

The standard method in its various forms is the most compact and does not require any drawing. The only point at which you have to worry about place value is at the start of each line of working and that is easy because you start immediately underneath the digit that you are multiplying. The drawback is that you do have to concentrate or you can easily lose track of where you are in the calculation. The most common reason for getting lost is not knowing the multiplication tables well enough – sorry to keep harping on about this, but it's true.

Trailing Zeroes
Trailing zeroes are zeroes that show the place value by filling up the blank digits to the right of a number. If you are multiplying numbers that have trailing zeroes, do not include them in the calculation or you will have a huge layout that goes all over the page and consists mostly of line of zeroes.

For example, if you are multiplying 45 000 x 17 300, you can adjust the place value by counting up the number of trailing zeroes in both numbers and then removing them. In this example there are 5 zeroes.

Now work out 45 x 173 by whatever method you use. Three methods are shown on the next page.

Table Method:

	100	70	3	
40	4000	2800	120	6920
5	500	350	15	865
				7785

Working:

4000	500
2800	350
120	15
6920	865

Lattice Method:

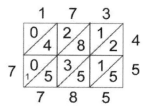

Standard Method:

Note that it has been done as 173 x 45 in order to have just two lines of working, otherwise there would be three.

$$
\begin{array}{r}
1\ 7\ 3 \\
4\ 5 \\
\hline
8^3\ 6^1\ 5 \\
6^2\ 9^1\ 2\ 0 \\
\hline
7\ 7\ 8\ 5 \\
\hline
1 \\
\end{array}
$$

So 45 x 173 = 7 785 (seven thousand, seven hundred and eighty five).

Now shift it this answer to its correct place value by putting back the 5 trailing zeroes:

45 000 x 17 300 = 778 500 000 (seven hundred and seventy five million, five hundred thousand).

Other Zeroes

If there are zeroes in amongst the non-zero digits, such as 4006 x 303, you can ignore them in the table method, but you must take them into account in the lattice and standard methods. Here is 4006 x 303 by all three methods:

Table Method:

	4000	6	
300	1 200 000	1 800	1 201 800
3	12 000	18	12 018
			1 213 818

There are only four non-zero digits and so there are only four multiplications, 3 x 4 (twice) and 3 x 6 (twice), but you must get the place values right. You do not need a row and two columns for the zeroes.

Lattice Method:

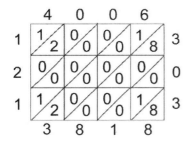

You **must** include a row and two columns for the zeroes. Most of the lattice then contains zeroes, but if you were to leave them out, the place values of all the non-zero entries would be wrong

Standard Method:

```
        4  0  0  6
           3  0  3
      1  2  0  1¹ 8
   1  2  0  1¹ 8  0  0
   1  2  1  3  8  1  8
```

The row of working for 3 x 4006 is easy but you must not forget to multiply 3 x 0 in the tens and hundreds place values or the place value for the second 3 x 4 will be wrong.
There is no point in putting in a line of working for multiplying by the 0 tens in 303 because it will all be zeroes and will contribute nothing to the answer.
The first line of working is then repeated for 300 x 4006, but it is shifted two place values to the left.

That completes multiplication.

Division

Division is repeated subtraction, which is how calculators do it. We are going to start with this simple but inefficient method and see how we can use place value and our knowledge of multiplication tables to speed it up.

Division methods are often called 'short' division, where working is mostly done in your head, and 'long division' where it is written out. Usually, short division is taught first using small numbers and then long division is taught later with larger numbers. I am going to turn this round and show you first how long division works and then how it can be reduced to the short method.

Here is the basic principle of division and the way that calculators do it. For example, $512 \div 32$.

5^4 ^{11} 2	1		Start by subtracting 32 from 512. Start a count of
3 2			how many times you do this subtraction.
4 8 0			

4 8^7 ^{10}	2	Do it again
3 2		
4 4 8		

4 4 8	3	And again.
3 2		
4 1 6		

If you keep on doing this (I don't recommend that you actually try it) the number 512 will be reduced to 32 and you will eventually get to this:

3 2	16	We have reached zero, so we can't subtract 32
3 2		again. It took 16 subtractions to do this, so that's
0		the answer. $512 \div 32 = 16$

This is the way that calculators do division.

If you ask a calculator to work out 10 000 000 by 4, it will subtract 4 from 10 000 000, giving 9 999 996, then subtract again, giving 9 999 992, and again, giving 9 999 988, and so on until it reaches 0. It keeps a count of how many times it does this subtraction, which is 2 500 000 times.

This is a horribly inefficient method but it is simple, requires no knowledge of multiplication tables, and the calculator does it with blinding speed.

It is not recommended for humans! We can use a method that exploits our knowledge of multiplication tables and this is shown below.

We can improve on this.

Let's do 512 ÷ 32 again. We can subtract several lots of 32 at a time, and although we could choose any amount of 32s, it makes sense to subtract hundreds or tens of 32s because we can easily calculate how many we are going to subtract.

First of all, consider subtracting hundreds of 32s. Can we subtract 100 x 32? No, because 100 x 32 = 3200 and we are trying to subtract from 512, so that would be way too many.

Now consider subtracting tens of 32s. What about 10 x 32? That's 320, so yes, we could do that. How about 20 x 32? No, because that makes 640 and that's too much.

So, let's subtract 10 x 32:

```
5⁴  ¹1  2
3   2   0          10 x 32
―――――――――
1   9   2
```

We have 192 left. So how many 32s can we take from that? It could be any number between 1 and 9. Let's try 5: 5 x 32 = 160. We could subtract that. How about 6? 6 x 32 = 192. Yes! It goes exactly 6 times:

Working:

```
    3   2                    3   2
        5                        6
―――――――――            ―――――――――
1   6   0            1   9   2
    1                    1
```

$$5^4\ ^1 1\ 2$$
$$\underline{3\ 2\ 0} \qquad 10 \times 32$$
$$1\ 9\ 2$$
$$\underline{1\ 9\ 2} \qquad 6 \times 32$$
$$0$$

So, adding together the 10 times and the 6 times that we took 32 away from 512, we can subtract 32 from 512 sixteen times. In other words, 512 ÷ 32 = 16.

This is an improvement on the calculator method. Firstly, we have reduced the number of subtractions from sixteen to two by subtracting multiples of 32 instead of doing it one at a time. Secondly, we have kept the required multiplications as simple as possible by working through the place values: hundreds, tens, then units.

Notice that we do need to work out at least some of the multiples of 32. There isn't any way to avoid it.

This is, in principle, how we do division. All we need do now is find a better way of laying it out than a series of separate subtractions.

Long Division

This uses exactly the same approach that we have just seen. We work our way through the place value columns, starting from the largest on the left, asking the question "What is the largest number that we can subtract?".

Here is 512 ÷ 32 again, using the conventional 'bus stop' layout for long division:

$$3\ 2\ \lfloor \overline{\mathbf{5}\ 1\ 2}$$

Start with the hundreds place value, which is a 5. Can we subtract 32 from 5? No. So there won't be any hundreds in the answer and the 5 is a remainder.

$$\qquad\qquad 1$$
$$3\ 2\ \lfloor \overline{\mathbf{5}^4\ ^1 1\ 2}$$
$$\qquad \underline{3\ 2}$$
$$\qquad 1\ 9$$

Now consider the tens. Together with the 5 hundreds we have 51 tens. Can we subtract 32 from 51? Yes. Could we subtract 2 x 32 which is 64? No. So, firstly write a 1 above the line in the tens place value. Next, subtract 32 from 51 in a subtraction written below the line. The answer to this subtraction gives the remainder: 19 tens.

```
            1 6
  3 2 |5⁴ ¹1 2
      3 2
      1 9 2
      1 9 2
          0
```

Now consider the units. We have 2 there but we must include the remainder of 19 tens to make 192 altogether. Do this by 'bringing down' the 2 and writing it next to the remainder 19.

How many 32s can we subtract from 192. The problem now is that we don't know the 32 times table. The only thing to do is to make an estimate and do some working separately, just as we did in the previous example. We might have to do several multiplications before we find the largest number of 32s we can take from 192. It turns out that 6 x 32 = 192, so it 'goes exactly' 6 times. Write 6 above the line and the subtraction below.

The answer: 512 ÷ 32 = 16. This example has been chosen so that it does go exactly, with no remainder. We will see what to do with remainders in later chapters, firstly by treating it as a fraction and, secondly, by using decimal places.

The procedure shown above works exactly the same for larger numbers. Here is 89241 ÷ 197 (a bit unlikely, true!) shown in steps including all working to get the multiples of 197:

```
  1 9 7 |8 9 2 4 1
```

Can you subtract 197 from 8? Obviously not.

What about 89? Again, no.

```
  1 9 7 |8 9 2 4 1
```

```
              4
  1 9 7 |8 9⁸ ¹2 4 1
        7 8 8
        1 0 4
```

Working
```
  1 9 7
      4
  7 8 8
  3 2

  1 9 7
      5
  9 8 5
  4 3
```

OK, what about 892? Yes, but how many times? 197 is close to 200 and that would go 4 times. So work out 4 x 197. It comes to 788. What about 5 times? That is 985 which is too many. So, it goes 4 times. Write 4 above the line and do the subtraction below.

Working
```
  1 9 7
        6
1 1 8 2
  5   4
```

Now consider the next place value. 'Bring down' the 4 so that our remainder is 1044. We already know that 5 x 197 is 985, so that would go. What about 6 times? That is 1182, which too many. So it goes 5 times. Write 5 above the line and do the subtraction below.

Working
```
  1 9 7
        3
  5 9 1
  2   2
```

Now consider the units. Bring down the 1 to make 591. Try 3 times 197, because we already know that 4 x 197 is 788 which is too big. It turns out that 3 x 197 = 591, which goes exactly. Write 3 above the line and do the subtraction below.

The answer: 89241 ÷ 197 = 453.

Short Division

Short division is exactly the same procedure as long division, except that you do all the subtracting in your head and carry the remainders, written small into the next place value.

Here are the two examples, 512 ÷ 32 and 89241 ÷ 197 written using short division layout next to their long division version. The remainders in the long divisions are shown in bold so you can see how they are the same as the carried remainders in the short divisions.

Long Division

$$3\,2\,|\,\overline{5^4\;{}^1 1\;2}$$

```
          1   6
3 2 | 5⁴  ¹1   2
      3   2
      1   9   2
      1   9   2
              0
```

Short Division

```
          1    6
3 2 | 5  ⁵1  ¹⁹2
```

Step 1: 32 won't go into 5, so carry the remainder 5 into the next column, making 51.

Step 2: 32 goes into 51 once, carry the remainder 19, making 192.

Step 3: 32 goes into 192 6 times exactly.

Long Division

```
              4   5   3
1 9 7 | 8  9⁸  ¹2   4   1
        7   6   8
        1⁰ ¹0⁹ 4¹³ ¹4
            9   8   5
            5   9   1
            5   9   1
                    0
```

Short Division

```
              4     5     3
1 9 7 | 8  9  ⁸⁹2  ¹⁰⁴4  ⁵⁹1
```

Step 1: 197 won't go into 8 or 89, so carry 89 into the next column, making 892.

Step 2: 197 goes into 892 4 times, carry the remainder 104, making 1044.

Step 3: 197 goes into 1044 5 times, carry the remainder 59, making 591.

Step 4: 197 goes into 591 3 times exactly.

The decision to use short division rather than long division really depends upon whether or not you think that you can do all the carrying and subtracting in your head on top of all the working that is needed to get the multiples of 32 or 197. As you can see, it is perhaps possible with $512 \div 32$ but, very hard with $89241 \div 197$.

Here is an example of a division which can quite easily be done with short division layout, $1424 \div 4$:

```
          3    5    6
4 | 1  ¹4  ²2  ²4
```

Step 1: 4 won't go into 1, carry the 1.
Step 2: 4 goes into 14 three times, carry the remainder 2.
Step 3: 4 goes into 22 five times, carry the remainder 2.
Step 4: 4 goes into 24 six times exactly.

I strongly recommend that you reserve the short division layout for divisions like 1424 ÷ 4 where you can easily do the 'divide, subtract, carry' procedure in your head; in other words, when you are dividing by a single digit number. For anything harder, use long division. Please. Or, reach for the calculator.

Checking the Answer

If you have done a division and you aren't sure that you have done it right, you can always check it by multiplying your answer by the divisor. In the last example:

```
  3 5 6
      4
  ‾‾‾‾‾
1 4 2 4
  2 2
```

If you get back the number you divided, 1424, the division must be correct.

Practice Exercises

You can download practice examples of the four rules of number from http://numbersexplained.co.uk.

Chapter 8 – Negative Numbers

In this chapter we will look again at negative numbers and see how to handle minus signs in arithmetic expressions.

We started out with **counting numbers**, also called **positive integers**. A good way of representing them is to write them on a line starting with zero and increasing to the right, like this:

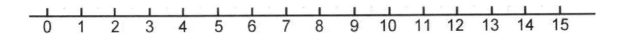

These numbers are positive. It is important to understand that all numbers have an operator. The operator is in front and tells you which way you are moving. The number itself then tells you how many steps. They can be written with a plus sign in front, for instance +5. This plus sign shows that it is located on the number line 5 steps up the number line from zero. Normally we don't bother to write the plus sign. If you see a number on its own, you assume that it is positive.

We have already seen that a subtraction such as 3 – 5 gives a number to the left of zero. 3 - 5 = -2. Whether this actually has any meaning or not depends upon the context. If we are talking about the number of people in a room, 3 – 5 is nonsensical. If we are talking about temperatures, 3 – 5 can mean that the temperature was 3° and it has dropped by 5° to -2°. To allow this calculation, we extend the number line to the left of zero to include negative numbers:

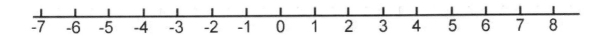

A negative number must have its minus sign written in front, like this: -3. This is located 3 steps back down the number line from zero.

The number line can be extended forever in both directions. It goes to infinity to the right and to minus infinity to the left. Infinity is not really a number but a way of saying that a procedure goes on forever. No matter how big a number you think of, you can always find a bigger one.

The Operators Plus and Minus

If we think of positive numbers and the plus sign as the natural way of counting, you can think of the plus sign as "keep going". Therefore 3 + 4 is + 3 + 4 and means "go three steps up the number line from zero and then keep going another four steps".

What the minus sign says is "go the other way". Instead of stepping up the number line to the right, go back down it to the left. So 6 − 2 is + 6 − 2 and means "go six steps up the number line from zero and then go back the other way two steps".

This is another way of saying that addition and subtractions are inverse operations. If you have 4 − 4 (which is really +4 − 4), then you end up back at 0 where you started — they cancel each other out.

Directed Numbers

We will now consider what happens if you combine these operators. What does it mean if a number has more than one operator in front? This is often called a 'directed number'.

If we look at the way we can combine two plus or minus signs, there are four possibilities: + and +, + and -, - and +, - and -. Here's what they mean:

	Meaning
+ +	Keep going, keep going
+ −	Keep going, go the other way
− +	Go the other way, keep going
− −	Go the other way, go the other way

There are only two possible outcomes to these instructions. You either go up the number line to the right or go back down it to the left. So what is the outcome for each of these four combinations of signs?

$+ +$ means "keep going, keep going". This just means "keep going", so that must mean plus.

$+ -$ means "keep going, go the other way". This means that you will end up going the other way, so that's minus. That's also true for $- +$ because the order of the instructions shouldn't matter.

$- -$ means "go the other way, go the other way". This is an instruction to turn around twice, so that you will be going up the number line the way you started. So $- -$ means plus.

Here is a summary:

	Meaning	Result
$+ +$	Keep going, keep going	$+$
$+ -$	Keep going, go the other way	$-$
$- +$	Go the other way, keep going	$-$
$- -$	Go the other way, go the other way	$+$

You may be wondering how a number can have two operators. In reality, it doesn't happen much in everyday arithmetic, but here is an example. Suppose that you wanted to calculate the temperature difference between 2°C and -3°C. That would be the arithmetic expression $2 - -3$. The rules say that $- -$ means +, so $2 - -3 = 2 + 3 = 5$. This is correct ; there are indeed 5 steps between 2 and -3, as shown below.

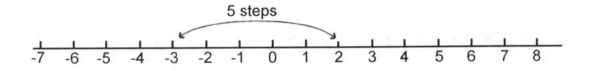

Like signs are plus, unlike signs are minus.

In the last example, the sign of the answer is positive, but the diagram showed the five steps between 2 and -3 as negative. This seems odd. Surely it should be negative? If you perform the calculation the other way round, - 3 - + 2, you do get a negative answer.

Here are the two calculations following the directed number rule 'like signs plus, unlike signs minus':

2 - - 3:

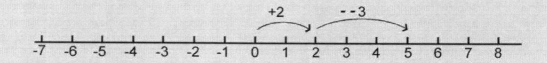

-3 - + 2:

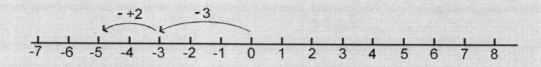

A difference calculation can result in either sign depending upon the way it is set up and it needs to be interpreted with caution.

Here are some other additions and subtractions of directed numbers so that you can see the 'like signs plus, unlike signs minus' rule at work:

$$2 + {}^{+}3 \ = \ 2 + 3 \ = \ 5$$

$$2 + {}^{-}3 \ = \ 2 - 3 \ = \ -1$$

$$2 - {}^{+}3 \ = \ 2 - 3 \ = \ -1$$

$$2 - {}^{-}3 \ = \ 2 + 3 \ = \ 5$$

Multiplication and Division of Directed Numbers

Two plus or minus operators can also be created by multiplying or dividing directed numbers. For example, 2 x -3. If you remember that the 2 has an unwritten + in front, all you have to do is put both the + from the 2 and the − from the 3 in front of the answer: +2 x -3 = +-6. As we have seen, + and − makes −, so the answer is -6. Here are the possibilities:

$$2 \times 3 \quad = \quad +2 \times +3 \quad = \quad ++ 6 \quad = \quad +6 = 6$$

$$2 \times -3 \quad = \quad +2 \times -3 \quad = \quad +-6 \quad = \quad -6$$

$$-2 \times 3 \quad = \quad -2 \times +3 \quad = \quad -+6 \quad = \quad -6$$

$$-2 \times -3 \quad = \quad --6 \quad = \quad +6 = 6$$

Division works in the same way:

$$8 \div 2 \quad = \quad +8 \div +2 \quad = \quad ++ 4 \quad = \quad +4 = 4$$

$$8 \div -2 \quad = \quad +8 \div -2 \quad = \quad +-4 \quad = \quad -4$$

$$-8 \div 2 \quad = \quad -8 \div +2 \quad = \quad -+4 \quad = \quad -4$$

$$-8 \div -2 \quad = \quad --4 \quad = \quad +4 = 4$$

Again, you can remember this as "like signs give plus, unlike signs give minus".

Directed Numbers With More Than Two Operators

If you can have two operators in front of a number, then why not have three or more? Well, you can and the rules aren't difficult if you think that + means "keep going" and minus means "go the other way". The result can only be either plus or minus and it will depend upon how many minus signs there are. If there are an even number of "go the other way" instructions, the result will be plus, a positive number. If there are an odd number of "go the other way" instructions, the result will be minus, a negative number. Here are some examples:

---3 = -3 because there are an odd number of minus signs

-+-3 = +3 = 3 because there are an even number of minus signs

You may be wondering how you can get three operators stacked up like this. A multiplication of three directed numbers would do it:

-4 x -5 x -2 = ---40 = -40

-4 x 5 x -2 =-4 x +5 x -2 = -+-40 = +40 = 40

> **An even number of minus signs gives plus, and an odd number of minus signs give minus.**

Directed numbers are not common in everyday arithmetic, but they do become important in algebra.

A Review of the Set of Integers

So far we have developed a set of numbers called integers. They start at zero. They include all the natural counting numbers, which are also called positive integers, and these go on forever to infinity. They also include the negative integers which go on forever towards minus infinity.

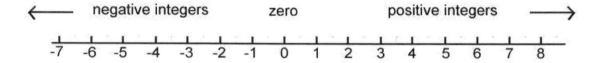

Note that zero is a special number: it has no sign and is neither positive nor negative.

We know how to add and subtract these numbers and how to do the repeated versions of these operations, multiplication and division.

We can write arithmetic expressions such as 4 + 3 x 2 and we have a set of rules that tell us to calculate their value. In most cases we can get an answer that will be another member of the set of integers. In the example 4 + 3 x 2, it is the number 10. (If you thought it was 14, go back and re-read chapter 6.) However, there is a problem with division: we don't always get an answer that is another integer. For example, 12 ÷ 5 = 2 remainder 2. We will deal with this difficulty when we extend our numbers to include fractions in the next chapter.

The set of integers goes in steps. For instance, there is no counting number between 2 and 3. The mathematical word for this is **discrete**. This means that we can count using integers, but that measuring things is a problem because they might not be an exact number of units. Again, this is solved by using fractions.

Thou Shalt Not Divide by Zero

Zero is a special number in another way, as stated before. You cannot divide by zero. The reason is simple: division is repeated subtraction. So if you try to divide a number by zero you are asking the question "How many times can I take zero away from this number?" and you can do that for ever. Try it on a calculator and will give you an error message.

Some people say that dividing by zero gives you infinity, which has some truth in it. However, infinity is not really a number but a signal that a process never stops or that some quantity has no bounds. It is better to say that the result of dividing by zero is undefined.

Here is another way of thinking about it. Differential calculus is an important branch of maths that considers the way that numerical systems can change. If we use the language of calculus, we can say that as a divisor 'tends towards' zero, the dividend 'tends towards' infinity.

Are addition and subtraction the same thing?

Every positive integer has a negative inverse. For example, 3 has the inverse -3 and 1 000 000 has -1 000 000.

Now that we seen the rules of directed numbers, it is possible to take the view that there is not really any such thing as subtraction, merely the addition of negative numbers. This means that, 7 – 4 is really +7 + -4.

When the topic of subtraction was introduced, I pointed out that whereas it is true that 3 + 4 can be written in the reverse order 4 + 3 and both expressions make 7, this is not true of subtraction. 7 – 4 is not the same as 4 – 7.

If we treat subtraction as merely adding directed numbers, this peculiarity goes away. +7 + -4 = +3 and also -4 + +7 = +3.

All this may be true, but it doesn't reflect well the everyday experience of 'taking away' physical objects, where one positive number of objects is subtracted from another and negative numbers don't seem to make any sense.

Let's keep subtraction.

Practice Exercises

You can download practice examples of arithmetic with directed numbers from http://numbersexplained.co.uk.

Chapter 9 – Fractions

First we will look at **closure** in sets of numbers. This was mentioned briefly in connection with the four rules of number. What does closure mean?

If you add any two counting numbers together, you always get another counting number. Since multiplication is repeated addition, multiplying two counting numbers will also result in another counting number. This is called closure.

You have also seen that the set of counting numbers is not closed for subtraction. This problem was solved by extending the set of counting numbers to include negative numbers. This new set of numbers is called the set of integers.

So, the set of integers is closed for addition, subtraction and multiplication.

Dividing Integers

There is a problem with division. Firstly, consider 20 ÷ 4.

Division is repeated subtraction. You can think of this in two ways. One way is to ask how many times you can subtract 4 from 20. It looks like this:

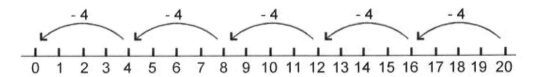

As the number line diagram shows, it can be done 5 times.

The other approach is to ask what is the largest number that can be subtracted from 20 four times. It looks like this:

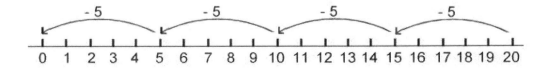

You can take away 5, four times exactly.

Either way, the division of these two counting numbers 20 and 4 gave us another counting number, 5. This is because 20 is a multiple of 4. But what happens if the number you are dividing isn't a multiple of the divisor and it doesn't divide exactly?

For example, 23 ÷ 4 doesn't go exactly. Here's what it looks like, using both approaches:

Firstly, how many times can 4 be subtracted from 23?

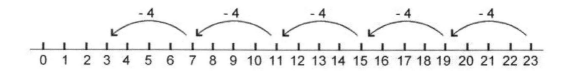

And secondly, what is the largest number that can be subtracted four times from 23?

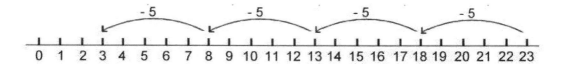

You can subtract 4 five times or subtract 5 four times, but there are three left over and we are forced to say that the answer to 23 ÷ 4 is "5 remainder 3". This is not a number but an admission of failure.

So, the set of integers is not closed for division.

We need some way of dividing the remainder of 3 by 4.

How to Divide Three by Four

Here is the number three laid out on a number line.

We want to start at 3 and take away some number four times, ending up exactly at 0. We can do it if we break up the interval between each integer into four equal parts, like this:

This new interval is a quarter. Now, working in quarters, we can subtract a number four times exactly:

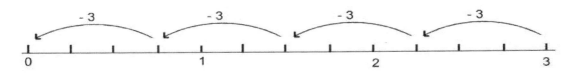

This number is 3, but be clear that it isn't the integer 3 because we aren't counting in integers now. It is 3, but counting in quarters.

We need is a way of noting that we have broken the counting number interval into four equal parts. We do this by drawing a short line under each number and writing a 4 to show that we are counting in quarters. This is a **fraction**.

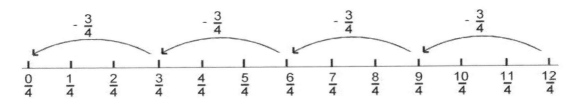

So, the answer to 3 ÷ 4 is the fraction three quarters, written $\frac{3}{4}$.

Mixed Numbers

Earlier in this chapter we considered 23 ÷ 4 and got the answer 5 remainder 3. We can now deal with that annoying remainder by breaking it down into quarters:

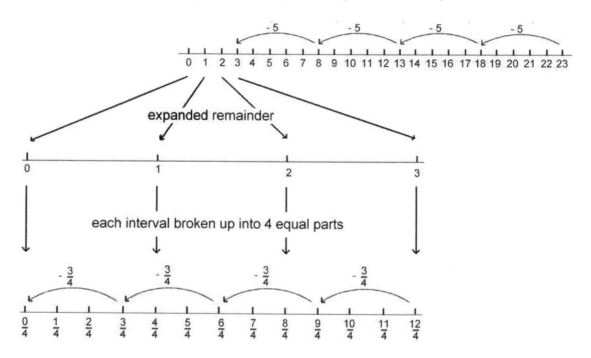

So, $23 \div 4 = 5\frac{3}{4}$

You may object that $5\frac{3}{4}$ is a pretty odd looking number. You would be right. It mixes up two different types of number, integers and fractions – indeed it is called a **mixed number**. Not only that, it contains a hidden addition because it means 5 and three-quarters: $5 + \frac{3}{4}$.

Why does it matter that $5\frac{3}{4}$ means $5 + \frac{3}{4}$? The reason is that it can get a bit messy if you want to add, subtract, multiply or divide mixed numbers.

For example, $5\frac{3}{4} \times 2\frac{1}{2}$ is actually $(5 + \frac{3}{4}) \times (2 + \frac{1}{2})$. This has to be multiplied out to give $5 \times 2 + 5 \times \frac{1}{2} + \frac{3}{4} \times 2 + \frac{3}{4} \times \frac{1}{2}$. The whole calculation looks like this:

$$5\frac{3}{4} \times 2\frac{1}{2}$$

$$= \left(5 + \frac{3}{4}\right) \times \left(2 + \frac{1}{2}\right)$$

$$= 5 \times (2 + \frac{1}{2}) + \frac{3}{4} \times (2 + \frac{1}{2})$$

$$= 5 \times 2 + 5 \times \frac{1}{2} + \frac{3}{4} \times 2 + \frac{3}{4} \times \frac{1}{2}$$

$$= 10 + 2\frac{1}{2} + 1\frac{1}{2} + \frac{3}{8} \qquad \text{(This addition will be explained later.)}$$

$$= 14\frac{3}{8} \qquad \text{Don't worry, there is a better way.}$$

The next chapter will sort this out by defining a new set of numbers that includes all the integers and all the fractions. This is the set of rational numbers.

Most of us were introduced to fractions in a slightly different way.

You may have been told that a fraction is an **instruction** to divide and multiply. For instance, $\frac{3}{4}$ of 20 means divide 20 by 4 into quarters and then take three of them. That is: divide by 4 and then multiply by 3, giving 15.

This is perfectly correct interpretation, but it doesn't take us very far. What do you do with $\frac{3}{4}$ of 18? There isn't an integer answer.

At some point, you have to get to grips with the idea of a fraction as a **number** in its own right, not just an instruction.

Chapter 10 – Rational Numbers

In this chapter, we will define a new set called **rational numbers** that includes all the integers and all the fractions, both positive and negative. This set is closed for all four operations + - x and ÷. This means that any operation on rational numbers will give you another rational number.

This set of numbers will also possess an important property that allows us to measure things and not just count them: they are **continuous**.

Rational numbers are often called just 'rationals'.

The Definition of a Rational

A rational is an ordered pair of integers. The usual way to write this pair of integers is one above the other with a line between – the way fractions are written.

$$\frac{\text{numerator}}{\text{denominator}}$$

The top one is a counting number called the **numerator** and can be any integer.

The bottom one is the **denominator** which can be any integer except 0. The denominator tells us what scale is being used when counting with the numerator. This needs some further explanation.

> A rational number is any pair of integer numerator and non-zero integer denominator. The numerator is the counting number and the denominator tells us what scale is being used.

The Meaning of the Denominator

With counting numbers, we jump from one number to the next, like walking across stepping stones. (The technical word for this is **discrete**.) There are gaps between each number.

With rationals, we fill in the gaps. The interval between each whole number is broken up into equal parts and more 'stepping stones' are provided so that you don't have to 'jump' so far.

The denominator tells us how many equal parts we have created when breaking up the whole number interval. If you like, it is how many extra stepping stones have been provided.

Writing Integers as Rational Numbers

The set of rationals includes all the fractions such as $\frac{1}{2}, \frac{1}{4}, \frac{2}{5}$, etc.. It also includes all the integers (counting numbers). These all have a denominator of 1.

For example, the integer 2 is equal to the rational number $\frac{2}{1}$.

The denominator of 1 is effectively saying that the counting interval is being left as a single whole step.

Fitting the Numbers Together

Here is a number line showing how integers and fractions fit into the set of rationals.

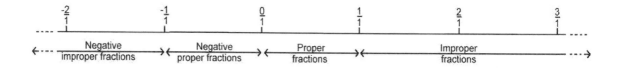

It also distinguishes **proper** fractions in the range between 0 and 1 and all the other fractions which are called either **improper** or 'top-heavy' fractions. Both proper and improper fractions also have their negative counterparts to the left of zero.

Some examples of rational numbers are given in the following number lines diagrams, showing how their positions relate to the integers.

Halves, with 2 steps to each integer:

Thirds, with 3 steps to each integer:

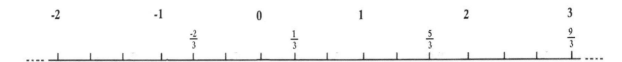

Fourths, with 4 steps to each integer:

Eighths, with 8 steps to each integer:

Who Uses Improper Fractions?

You may object that in everyday life we use mixed numbers, not improper fractions. We don't talk about $\frac{5}{2}$ or five halves but use the mixed number $2\frac{1}{2}$.

That is true, but $2\frac{1}{2}$ is actually not one number but a little addition sum: $2 + \frac{1}{2}$. This makes it tricky to use in calculations. Mixed number calculations are explained in a later chapter and you will see that it is often easier to convert mixed numbers into improper fractions, perform the calculation and then convert back to a mixed number.

Equivalence Sets of Rationals

This section looks at an important and most remarkable property of rationals.

Suppose that you were sharing a cake equally with a friend. The friend cuts the cake into four equal pieces and gives you two of them and takes the other two for himself. You might be puzzled, but you wouldn't feel cheated – you have been given half of the cake.

The same would be true if he cut the cake into six pieces and gave you three, or if he cut it into eight and gave you four.

Using the notation of rationals, this is written:

$$\frac{1}{2} = \frac{2}{4} = \frac{3}{6} = \frac{4}{8}$$

Notice how the numbers are connected:

$$\overset{\times 2}{\frac{1}{2}} = \underset{\times 2}{\frac{2}{4}} = \frac{3}{6} = \frac{4}{8} \quad \text{and} \quad \overset{\times 3}{\frac{1}{2}} = \frac{2}{4} = \underset{\times 3}{\frac{3}{6}} = \frac{4}{8} \quad \text{and} \quad \overset{\times 4}{\frac{1}{2}} = \frac{2}{4} = \frac{3}{6} = \underset{\times 4}{\frac{4}{8}}$$

Starting from $\frac{1}{2}$, we can get $\frac{2}{4}$ by multiplying numerator and denominator by 2, and we can get $\frac{3}{6}$ by multiplying them by 3, and $\frac{4}{8}$ if we multiply them by 4.

Also, we don't have to start with $\frac{1}{2}$:

$$\frac{1}{2} = \overset{\times 2}{\underset{\times 2}{\frac{2}{4}}} = \frac{3}{6} = \frac{4}{8}$$

In fact, we can multiply the numerator and denominator of any rational number by **any** integer and get another equal rational number. Here are some examples:

$$\overset{\times 2}{\underset{\times 2}{\frac{3}{4}}} = \frac{6}{8} \qquad \overset{\times 4}{\underset{\times 4}{\frac{5}{8}}} = \frac{20}{32} \qquad \overset{\times 100}{\underset{\times 100}{\frac{2}{3}}} = \frac{200}{300} \qquad \overset{\times 5}{\underset{\times 5}{\frac{3}{16}}} = \frac{15}{80} \qquad \overset{\times 6}{\underset{\times 6}{\frac{7}{12}}} = \frac{42}{72}$$

We can also **divide** the numerator and denominator of any rational number by any integer factor and get another equal rational number. Here are some examples:

$$\frac{12}{24} \overset{\div\,12}{\underset{\div\,12}{=}} \frac{1}{2} \qquad \frac{20}{30} \overset{\div\,10}{\underset{\div\,10}{=}} \frac{2}{3} \qquad \frac{54}{72} \overset{\div\,18}{\underset{\div\,18}{=}} \frac{3}{4} \qquad \frac{250}{1000} \overset{\div\,250}{\underset{\div\,250}{=}} \frac{1}{4} \qquad \frac{35}{80} \overset{\div\,5}{\underset{\div\,5}{=}} \frac{7}{16}$$

Here is a rule that you must understand and remember if you are going to perform calculations with rationals:

> **To change a rational number into another equal rational number, multiply or divide both numerator and denominator by the same integer.**

Now pause to consider:

Since we can multiply numerator and denominator by any integer and there are an infinite number of integers, this leads to a remarkable conclusion. It means that any rational can be written down in an infinite number of ways.

In other words, any rational number is a member of an infinite set of equal numbers. This set is called its **equivalence set**.

This is where rationals differ sharply from integers, which are all unique. For instance , there is only one integer number 5. However, if we convert 5 to a rational number $\frac{5}{1}$, then $\frac{5}{1} = \frac{10}{2} \quad \frac{15}{3} = \frac{20}{4}$ and so on infinitely.

Note that this is <u>not</u> saying that all rationals are equal to each other. For any rational, as well as there being an infinite number of rationals that <u>are</u> equal to it (its equivalence set), there are an infinite number of rationals that are <u>not</u> equal to it.

Lowest Terms

If you were asking directions from someone and they told you that the place you were looking for was 'four eighths of a mile' away, you would wonder why they didn't just say 'half a mile'. In other words, we usually use the first member of a rational number's equivalence set. This is the one that uses the smallest possible numbers for its numerator and denominator – we say that it is in its **lowest terms**.

To reduce a rational to its lowest terms, we need to look for numbers that will divide into both numerator and denominator. In other words, we need to divide them by any common factors.

For example, $\frac{24}{36}$ is not in its lowest terms. There is a common factor of 12.

$$\frac{24}{36} = \frac{2}{3} \quad \text{or, in two steps:} \quad \frac{24}{36} = \frac{6}{9} = \frac{2}{3}$$

A rational number is in its lowest terms if there are no common factors (other than 1) between the numerator and denominator.

The Continuous Nature of Rationals

We can make the denominator as large as we like. This means that we can count in steps that are as tiny as we like.

This allows us to measure things and consider the size of objects, or their weight (unlike integers that only let us count things and state how many objects there are). We can do this as accurately as we wish. If tenths of an inch are not good enough, then we can use hundredths. If that's not good enough, then use thousandths.

There is no limit to how fine this hair-splitting can be. If you take two rationals that are very close together, you can always find some more in between them.

For example, take $\frac{1}{1\,000\,000}$ and $\frac{2}{1\,000\,000}$ which are pretty close together. If we increase the denominator to 10 000 000, they become $\frac{10}{10\,000\,000}$ and $\frac{20}{10\,000\,000}$ and now there are the numbers $\frac{11}{10\,000\,000}$, $\frac{12}{10\,000\,000}$, $\frac{13}{10\,000\,000}$ and so on up to $\frac{19}{10\,000\,000}$ in between.

Rational numbers are continuous because we can make their denominators as large as we please. We can use this property to make measurements as accurately as we need.

The Importance of Rationals

The invention of rationals was an important step in the development of mathematics. You can perform most calculations with them. Rationals permit measurements and it is hard to see how science and engineering could have progressed without them.

The preceding paragraph talked about the 'invention of rational numbers'. However, are mathematical ideas invented or discovered? I could have written 'the discovery of rational numbers' and I nearly did.

Did rational numbers exist in some sense before the first mathematicians? If the world and the human species and all its works were destroyed tomorrow, would rational numbers and all other mathematical concepts, such as the circle, still exist? You decide.

This chapter has introduced a lot of ideas that take some digesting. Do not go on to the next chapters until you are confident that you have understood them.

Practice Exercises

You can download practice examples of changing rational numbers into other members of their equivalence sets, and of reducing them to lowest terms, from http://numbersexplained.co.uk.

Chapter 11 – The 'Four Rules' for Rational Numbers

This chapter sets out the rules for adding, subtracting, multiplying and dividing rational numbers. Some people would argue that you don't really need to know these procedures because they aren't used much in daily life. I would urge you to read through and try to understand them, even if you don't attempt to memorise them, because they shed some light on decimals. However, if you intend to do algebra later, you <u>must</u> learn them.

Addition

Remember that the counting part of a rational number is the numerator, the top number. The denominator tells us how many equal parts each whole number has been split into – if you like, what scale we are counting in.

If both numbers have the same denominator, there is no problem, we can just add the numerators. For example:

$$\frac{1}{5} + \frac{2}{5} = \frac{3}{5}$$

This also works fine for integers written as rationals:

$$\frac{2}{1} + \frac{3}{1} = \frac{5}{1}$$

Note that we do **not** add the denominators. This would make no sense because (it is worth repeating) they are not counting numbers but labels that tell us what scale we are using.

The answer might not be in lowest terms, so we might have to reduce to lowest terms after adding:

$$\frac{3}{8} + \frac{1}{8} = \frac{4}{8} \quad \overset{\div 4}{=} \quad \frac{1}{2}$$
$$\div 4$$

Otherwise, adding fractions with the same denominators is straightforward addition of the numerators.

Sadly, the denominators are often different. However, we can always change the fractions into other members of their equivalence sets so that the denominators are the same. In other words, we have to find a common denominator.

(If you don't know how to change a fraction into another member of its equivalence set, go back and review the last chapter.)

Here is a simple example. Some people (including me) prefer the layout on the right. It really doesn't matter:

$$\frac{1}{4} + \frac{1}{2}$$

$$= \frac{1}{4} + \frac{2}{4}$$

$$= \frac{3}{4}$$

$\frac{1}{2}$ has been changed into $\frac{2}{4}$ to get a common denominator.

$$\frac{1}{4} + \frac{1}{2}$$

$$= \frac{1 + 2}{4}$$

$$= \frac{3}{4}$$

In above example we had to change only one of the denominators, but it is often necessary to change both of them. For example (showing both layouts):

$$\frac{1}{5} + \frac{1}{2}$$

$$= \frac{2}{10} + \frac{5}{10}$$

$$= \frac{7}{10}$$

$\frac{1}{5}$ has become $\frac{2}{10}$ and $\frac{1}{2}$ has become $\frac{5}{10}$ to get a common denominator.

$$\frac{1}{5} + \frac{1}{2}$$

$$= \frac{2 + 5}{10}$$

$$= \frac{7}{10}$$

The common denominator must be a multiple of both the original denominators, but don't fall into the trap of just multiplying them together. That will work, but it might not be the best choice, as shown in the following example:

$$\frac{1}{8} + \frac{5}{12}$$

$$= \frac{12}{96} + \frac{40}{96}$$

$$= \frac{52}{96}$$

$$= \frac{13}{24} \text{ (in its lowest terms)}$$

96 is not the lowest common denominator. The first number that is both a multiple of 8 and a multiple of 12 is 24, as shown on the right.

$$\frac{1}{8} + \frac{5}{12}$$

$$= \frac{3}{24} + \frac{10}{24}$$

$$= \frac{13}{24}$$

Try to Use Lowest Common Denominators

Some students persist in always multiplying the denominators together to get a common denominator. If you do this, you will often end up working with much larger numbers than necessary and only have to reduce the answer down to its lowest terms at the end. This is particularly true if there are more than two rational numbers to add together. Compare these two possible ways of working the following example of adding four fractions:

$$\frac{1}{8} + \frac{1}{12} + \frac{1}{2} + \frac{1}{4}$$

$$= \frac{96}{768} + \frac{64}{768} + \frac{384}{768} + \frac{192}{768}$$

$$= \frac{736}{768} \quad \text{(common factor 2)}$$

$$= \frac{368}{384} \quad \text{(common factor 4)}$$

$$= \frac{92}{96} \quad \text{(common factor 4)}$$

$$= \frac{23}{24}$$

On the left, the denominators have been multiplied together, giving 768. Then the answer has been reduced to lowest terms using the common factors shown. On the right, the lowest common multiple of 24 has been used. Which is easier?

$$\frac{1}{8} + \frac{1}{12} + \frac{1}{2} + \frac{1}{4}$$

$$= \frac{3}{24} + \frac{2}{24} + \frac{12}{24} + \frac{6}{24}$$

$$= \frac{23}{24}$$

Do You Know Your Tables?

The above examples also demonstrate a fact that has been mentioned before: it is very difficult to do calculations with rational numbers unless you know multiplication tables very well. You literally need to know them backwards in order to spot common factors.

Subtraction

Subtraction is much like addition, except that you subtract the numerators. For example:

$$\frac{3}{4} - \frac{1}{2} = \frac{3-2}{4} = \frac{1}{4}$$

You can mix up additions and subtractions:

$$\frac{5}{8} + \frac{1}{4} - \frac{1}{3} = \frac{15+6-8}{24} = \frac{13}{24}$$

You can, of course, get a negative answer:

$$\frac{3}{5} - \frac{3}{4} = \frac{12-15}{20} = -\frac{3}{20}$$

That's it for addition and subtraction of rationals.

To add or subtract rational numbers, convert them to a common denominator (unless they already have the same denominator). This is the denominator of the answer. Then add or subtract the numerators.

Multiplication

Multiplication of rational numbers does what you might expect: you multiply the numerators together and you also multiply the denominators together. For example:

$$\frac{1}{2} \times \frac{3}{4} = \frac{1 \times 3}{2 \times 4} = \frac{3}{8}$$

You can see that this works for integers when written as rationals:

$$\frac{5}{1} \times \frac{3}{1} = \frac{15}{1}$$

You can also multiply more than one number, for instance:

$$\frac{1}{2} \times \frac{1}{2} \times \frac{1}{2} = \frac{1}{8}$$

Why do we multiply the denominators? After all, I said previously that for addition and subtraction, adding them makes no sense because they are just labels. Here is an explanation that I hope will satisfy you. The following diagram show a piece of paper representing 1 (or $\frac{1}{1}$). It shows one half shaded. $\frac{1}{2}$ means divide into 2 equal parts and select 1 of them.

Now we will take three quarters of this and show the result in a different shade. $\frac{3}{4}$ means divide a whole one into 4 equal parts and select 3 of them:

The whole sheet of paper is now divided into 8 equal parts and there are 3 doubly shaded parts representing the answer to $\frac{1}{2} \times \frac{3}{4} = \frac{3}{8}$. Note that the fractions had denominators of 2 and 4, meaning divide by 2 and also by 4. So the result must be a fraction where the sheet of paper is divided by 2 x 4 which is 8. That's why the denominators are multiplied when you multiply rationals.

To multiply two rational numbers, multiply the numerators and multiply the denominators. Then check that the answer is in its lowest terms.

Cancelling Out Common Factors First

It is not unusual for a multiplication to give an answer that isn't in its lowest terms. Dividing both numerator and denominator is called 'cancelling out common factors'. For example:

$$\frac{1}{3} \times \frac{3}{4}$$

$= \quad \frac{3}{12}$ There is a common factor of 3.

$= \quad \frac{\overset{}{\cancel{3}}\,1}{\cancel{12}\,4}$ When you cancel common factors, cross out the original numbers and write the new ones. You will see why soon.

$= \quad \frac{1}{4}$

Now look at above example again. You can see why there was a common factor of 3 in the answer. It was because there was a 3 in the numerator of one number and the denominator of the other. Why not cancel out this common factor before doing the multiplication?

Here is the same example:

$$\frac{1}{\cancel{3}\,1} \times \frac{\cancel{3}\,1}{4}$$ There is a common factor of 3 between the denominator of the first number and the numerator of the second.

$$= \frac{1 \times 1}{1 \times 4}$$ Multiply. (Usually you would do this line in your head)

$$= \frac{1}{4}$$ Now the answer is in its lowest terms straight away.

Always cancel out common factors between any numerator and any denominator before multiplying. You might think that this is an unnecessary complication in what is an otherwise straightforward operation, but it keeps the numbers small and makes it easy to spot the common factors. This is particularly useful if you are multiplying several numbers. Consider this example:

$$\frac{3}{4} \times \frac{5}{12} \times \frac{4}{5}$$ Suppose we ignore the common factors and just multiply it out.

$$= \frac{60}{240}$$ This is clearly not in its lowest terms. Cancel the obvious common factor of 10.

$$= \frac{6}{24}$$ Cancel out a common factor of 6

$$= \frac{1}{4}$$

Here it is again, this time cancelling out all the common factors first. This is shown in three steps for clarity, but you would normally do all three cancellations of common factors on one line of the calculation:

$$\frac{3}{4} \times \frac{\cancel{5}\,1}{12} \times \frac{4}{\cancel{1}\,\cancel{5}}$$ Cancel out the 5s immediately.

$$= \frac{3}{\cancel{4}\,1} \times \frac{\cancel{5}\,1}{12} \times \frac{1\,\cancel{4}}{\cancel{1}\,\cancel{5}}$$ Now the 4s

$$= \frac{1\,\cancel{3}}{\cancel{4}\,1} \times \frac{\cancel{5}\,1}{\cancel{4}\,\cancel{12}} \times \frac{1\,\cancel{4}}{\cancel{1}\,\cancel{5}}$$ Finally cancel out a common factor of 3. Then multiply.

$$= \frac{1 \times 1 \times 1}{1 \times 4 \times 1}$$ (Usually you would do this line in your head)

$$= \frac{1}{4}$$

Numbers Explained

It isn't a disaster if you don't spot all the common factors before multiplying – you will just get an answer that needs to be reduced to its lowest terms. However, it is good practice because it keeps the numbers small and so reduces the chance of error.

Division

Consider the integer division $3 \div 3$. The answer is 1. Let's write this in rational numbers. It must also be true that:

$$\frac{3}{1} \div \frac{3}{1} = \frac{1}{1} = 1$$

But here is a multiplication of $\frac{3}{1}$ by another rational number that also gives an answer of 1:

$$\frac{3}{1} \times \frac{1}{3} = \frac{3}{3} = \frac{1}{1} = 1 \qquad \text{or, better} \qquad \frac{\cancel{3}1}{1} \times \frac{1}{1\cancel{3}} = \frac{1}{1} = 1$$

So, it seems that $\times \frac{1}{3}$ has the same effect as $\div \frac{3}{1}$. In other words, multiplying by a third is the same as dividing by three. That isn't really a surprise, because taking a third of something means divide it into three equal parts.

What about $\div \frac{1}{3}$ and $\times \frac{3}{1}$? Let's try them out on some number, for example 2 or $\frac{2}{1}$. We know how to do $\times \frac{3}{1}$:

$$\frac{2}{1} \times \frac{3}{1} = \frac{6}{1} \text{ So the answer is 6.}$$

Is this what we would expect for $\frac{2}{1} \div \frac{1}{3}$? Well, there are 3 thirds in 1, so there would be 6 thirds in 2. So it does work.

What about if we divide a fraction by a fraction? If we divide $\frac{1}{2}$ by $\frac{1}{4}$ we would expect the answer 2, because 2 quarters make a half. So let's try $\frac{1}{2} \times \frac{4}{1}$:

$$\frac{1}{\cancel{2}1} \times \frac{2\cancel{4}}{1} = \frac{2}{1}$$

It works.

> **To divide by a rational number, multiply by its inverse. The inverse is the number 'turned upside-down'.**

> Previously, I asked if there was really a distinct operation called subtraction and suggested that it was merely the addition of a negative number.
>
> Now I am going to suggest that there is not really a distinct operation called division. Division is merely multiplication by an inverse. Arguably, this is true if we are working with rational numbers, since every number has an inverse.
>
> However, if we are working with integers and the numbers work out nicely, there is no need to employ rational numbers and we will let division survive.

The Importance of Rational Numbers

Rational numbers are all the possible ways that you can write a pair of integers as a $\frac{numerator}{denominator}$. There are many reasons why they are an important development in maths.

Firstly, we can represent almost any number as a rational. This includes all the integers (counting numbers), and all fractions including both proper fractions (between 0 and 1) and improper fractions (larger than 1). All these numbers also have their negative counterparts

The second reason is that rationals are a closed set for all the four rules of number. This means that you can add, subtract, multiply or divide any two rational numbers and you will get a rational number answer. There are also rules for mixing up these operations. There is just one exception: you cannot divide by 0.

In short, the set of rationals are well behaved numbers that will do almost anything that you might need in arithmetic.

There is a third reason. Rational numbers allow us to measure things with whatever accuracy we need because they are continuous. No matter how close together two rational numbers may be, we can always find some more in between, simply by increasing the denominators.

For these reasons, rational numbers allow the development of science and engineering.

In the 5th Century BC, the mystical cult of the followers of Pythagoras on the Greek island of Samos studied mathematics.

They believed that rational numbers had both scientific and religious significance. They observed that musical harmony is dependent on rational numbers and believed that harmony is a principle giving a structure to the universe and all aspects of reality. They thought that the motion of the planets depended on rational numbers and called it "the music of the spheres". This may sound quaint and naïve, but some modern theoretical physics uses much more complex mathematics to describe remarkably similar ideas concerning fundamental particles.

The Pythagoreans regarded their knowledge as sacred and only to be revealed to trusted members of the cult. They recognised that mathematics is the key to science and that engineering, which is applied science, is the source of economic and military power.

They also knew that there were some difficulties with their worship of rational numbers. For instance, if you consider a square that has sides of length one unit, there is no rational number that can give the exact length of its diagonal. If you believe that rational numbers rule everything, then that is a nasty problem. We will look at irrational numbers in Part 2.

Rational Number Expressions

The way rational numbers are written is often used in arithmetic expressions to show division. For example, you can write $\frac{12}{3}$ instead of $12 \div 3$.

This layout is often used instead of brackets to show the order of operations if there is a division involved. For example:

$2 + 4 \div 3$ can be written $2 + \dfrac{4}{3}$

$(2 + 4) \div 3$ can be written $\dfrac{2+4}{3}$

The line between numerator and denominator acts like a bracket.

This is often used to show repeated divisions without ambiguity:

$$\frac{\frac{16}{4}}{2} = \frac{4}{2} = 2 \qquad \text{but} \qquad \frac{16}{\frac{4}{2}} = \frac{16}{2} = 8$$

Note that the numbers are of different sizes and that the lines are of different length.

Calculators and Spreadsheets

Rational number layout is a convenient way of writing expressions, but you need to be careful when using a programmable calculator or spreadsheet. They are not always capable of reading arithmetic expressions correctly and you may need to use lots of brackets. For example, if you try to enter: $\dfrac{24}{2 \times 4}$, which is $\dfrac{24}{8} = \dfrac{3}{1} = 3$, you will probably get the answer to $\dfrac{24}{2} \times 4$, which is $\dfrac{12}{1} \times 4 = 12 \times 4 = 48$.

If you enter the wrong expression, you will get the wrong answer. Play safe and use brackets like this: $\dfrac{24}{(2 \times 4)}$.

Not all calculators would let you enter the expression like this anyway. You would have to enter $24 \div (2 \times 4)$. You need to learn how your particular calculator reads expressions. Some are more sophisticated than others.

As computer scientists say: garbage in, garbage out.

Practice Exercises

You can download practice examples of the four rules of number with rational numbers from http://numbersexplained.co.uk.

Chapter 12 – Mixed Numbers

We tend not to use improper fractions in everyday life. If you asked someone how far it was to a town and they replied that it was "fifteen over two miles" instead of "seven and a half miles", you would be a bit puzzled, even though the two answers are precisely equal.

$7\frac{1}{2}$ is an example of a mixed number. It is mixed because we are counting in integers as far as possible and then switching to a fraction. Here it is compared with $\frac{15}{2}$ on number lines:

Changing a Mixed Number into an Improper Fraction

To change a mixed number into an improper fraction, the integer part has to have its denominator made the same as the fraction part. Since its denominator is 1, that amounts to multiplying numerator and denominator by the denominator of the fraction. Then add the two numerators together. Here is the procedure explained step by step for $5\frac{3}{4}$:

$$5 + \frac{3}{4}$$ (A mixed number is actually a little addition sum.)

$$= \frac{5}{1} + \frac{3}{4}$$ (The integer part has a denominator of 1.)

$$= \frac{5 \times 4}{1 \times 4} + \frac{3}{4}$$ (Changing to a common denominator. This will always be the denominator of the fraction part.)

$$= \frac{20}{4} + \frac{3}{4}$$ (Now do the addition.)

$$= \frac{23}{4}$$ (Here's the answer.)

The above explains what is going on, but it boils down into a very simple procedure that does it all very quickly. Here it is for $5\frac{3}{4}$:

1. Multiply the integer by the denominator of the fraction: 4 x 5 = 20 .
2. Add the numerator of the fraction: 20 + 3 = 23 .
3. Write this answer as the numerator over the denominator of the fraction: $\frac{23}{4}$.

Changing an Improper Fraction into a Mixed Number

To change an improper fraction to a mixed number, the procedure is the reverse. Here it is for $\frac{23}{4}$:

1. Divide the numerator by the denominator: 23 ÷ 4 = 5 remainder 3 .
2. The integer part is the dividend: 5 .
3. The fraction part uses the remainder 3 as its numerator and keeps the denominator: $\frac{3}{4}$.
4. Write the integer and fraction part together: $5\frac{3}{4}$.

Here are some examples of mixed numbers and their improper fraction equivalents:

$$1\frac{1}{2} \quad = \quad \frac{3}{2}$$

$$10\frac{7}{8} \quad = \quad \frac{87}{8}$$

$$8\frac{11}{16} \quad = \quad \frac{139}{16}$$

The last example involves some fairly large numbers and you might not be able to do the multiplication (or division) in your head, but the procedure is the same however large the numbers.

The rest of this chapter goes on to explain how the 'four rules' work with mixed numbers. Mixed number calculations can be quite complicated and can easily require a lot of working. It is generally best to avoid them by using decimals unless they are obviously going to work out easily. You may decide to skip ahead to the next chapter on decimals.

Addition of Mixed Numbers

Addition of mixed numbers is quite straightforward because mixed numbers are themselves addition sums and you can change the order of an addition. Here is a simple example showing the logic of the procedure in full:

$$2\frac{1}{2} + 1\frac{1}{4}$$

$=$	$2 + \frac{1}{2} + 1 + \frac{1}{4}$	(Showing the unwritten additions in mixed numbers.)
$=$	$2 + 1 + \frac{1}{2} + \frac{1}{4}$	(Rearranging the order of the additions.)
$=$	$3 + \frac{2+1}{4}$	(Adding the integers and adding the fractions.)
$=$	$3 + \frac{3}{4}$	(The result is another mixed number.)
$=$	$3\frac{3}{4}$	

You do not normally write all this. This particular example you could do in your head but, here is a more normal layout for a mixed number addition:

$$2\frac{1}{2} + 1\frac{1}{4}$$

$=$	$3 + \frac{2+1}{4}$	(Adding the integers and adding the fractions.)
$=$	$3\frac{3}{4}$	

So, the rule is: add the integer parts together and add the fraction parts together.

You can add more than one mixed number together. Here is an example:

$$5\frac{1}{4} + 2\frac{1}{5} + 3\frac{3}{8}$$

$=$	$10 + \frac{10+8+15}{40}$	(Adding the integers and adding the fractions.)
$=$	$10\frac{33}{40}$	

There is one possible complication. The fraction part of the addition might give an improper fraction. The following example shows what to do.

$$5\frac{3}{4} + 2\frac{2}{5}$$

$$= \quad 7 + \frac{15+8}{20} \quad \text{(Adding the integers and adding the fractions.)}$$

$$= \quad 7 + \frac{23}{20} \quad (\frac{23}{20} \text{ is an improper fraction.)}$$

$$= \quad 7 + 1\frac{3}{20} \quad \text{(Converting the improper fraction to a mixed number.)}$$

$$= \quad 8\frac{3}{20} \quad \text{(Adding the integer parts together.)}$$

Subtraction

Subtraction of mixed numbers is very similar to addition. Here is an example showing the logic in full:

$$4\frac{3}{4} - 1\frac{1}{2}$$

$$= \quad 4 + \frac{3}{4} - (1 + \frac{1}{2}) \quad \text{(Each mixed number contains an addition.)}$$

$$= \quad 4 + \frac{3}{4} - 1 - \frac{1}{2} \quad \text{(That means subtracting 1 \underline{and} subtracting } \frac{1}{2} \text{.)}$$

$$= \quad 4 - 1 + \frac{3}{4} - \frac{1}{2} \quad \text{(Rearranging the order.)}$$

$$= \quad 3 + \frac{3-2}{4} \quad \text{(Subtracting the integers and subtracting the fractions.)}$$

$$= \quad 3 + \frac{1}{4}$$

$$= \quad 3\frac{1}{4} \quad \text{(Rewritten as a mixed number.)}$$

You do not normally write all this. This particular example you could do in your head but here is a more normal layout for the above mixed number subtraction:

$$4\tfrac{3}{4} - 1\tfrac{1}{2}$$

$$= \quad 3 + \frac{3-2}{4} \qquad\qquad \text{(Subtracting the integers and subtracting the fractions.)}$$

$$= \quad 3\tfrac{1}{4}$$

You can have more than one subtraction, but as with any repeated subtraction, the order that you do the subtractions gives a different answer. For example, $5\tfrac{1}{2} - 2\tfrac{1}{4} - 1\tfrac{1}{4}$ could mean $(5\tfrac{1}{2} - 2\tfrac{1}{4}) - 1\tfrac{1}{4}$ which is better written as $5\tfrac{1}{2} - (2\tfrac{1}{4} + 1\tfrac{1}{4})$. Either expression gives the answer 2. Or it could mean $5\tfrac{1}{2} - (2\tfrac{1}{4} - 1\tfrac{1}{4})$, which gives the answer $4\tfrac{1}{2}$.

There is a slight complication if the fraction part of the subtraction gives a negative answer. This is easily sorted out by taking 1 from the integer and changing it into a fraction. Here is an example:

$$4\tfrac{1}{2} - 1\tfrac{3}{4}$$

$$= \quad 3 + \frac{2-3}{4} \qquad\qquad \text{(The fraction subtraction will give } \tfrac{-1}{4} \text{) .}$$

$$= \quad 2 + \frac{4+2-3}{4} \qquad\qquad \text{(1 has been taken from the integers and added to the fractions as } \tfrac{4}{4} \text{.)}$$

$$= \quad 2\tfrac{3}{4}$$

Mixing Up Addition and Subtraction

As with integers and rational numbers, you can do additions and subtractions of mixed numbers together.

Here is an example:

$$3\tfrac{2}{3} + 4\tfrac{1}{4} - 1\tfrac{1}{2}$$

$$= \quad 6 + \frac{8+3-6}{12} \qquad \text{(3 + 4 − 1 gives 6. The fraction part will be done in } \tfrac{1}{12}\text{ths}$$
$$\text{and } \tfrac{8}{12} + \tfrac{3}{12} - \tfrac{6}{12} \text{ gives } \tfrac{5}{12}.)$$

$$= \quad 6\tfrac{5}{12}$$

Addition and subtraction of mixed numbers is done by adding or subtracting the integer and fraction parts separately and then combining the answers to give another mixed number.

Do <u>not</u> convert mixed numbers to improper fractions in order to add or subtract. It might look simpler, but you will probably have to work with large numerators and it will be tiresome to convert it back to a mixed number answer.

Multiplication

Multiplication looks as if you might just multiply the integer parts and multiply the fraction parts, but that doesn't work. That's because of the unwritten additions in each mixed number. Here is what you would really have to do if you wanted to multiply mixed numbers (not recommended). It breaks down into four separate multiplications:

$$2\tfrac{1}{2} \times 1\tfrac{2}{5}$$

$$= \quad (2 + \tfrac{1}{2}) \times (1 + \tfrac{2}{5}) \qquad \text{(Each mixed number is a little addition.)}$$

$$= \quad 2 \times (1 + \tfrac{2}{5}) + \tfrac{1}{2} \times (1 + \tfrac{2}{5}) \qquad \text{(Expanding the first bracketed sum.)}$$

$$= \quad 2 \times 1 + 2 \times \tfrac{2}{5} + \tfrac{1}{2} \times 1 + \tfrac{1}{2} \times \tfrac{2}{5} \qquad \text{(Expanding the second bracketed sum to give four multiplications added together.)}$$

$=$ $\quad 2 \times 1 + \frac{2}{1} \times \frac{2}{5} + \frac{1}{2} \times \frac{1}{1} + \frac{1}{2} \times \frac{2}{5}$ (Do each multiplication.)

$=$ $\quad 2 + \frac{4}{5} + \frac{1}{2} + \frac{2}{10}$ (Now we have four numbers to add together.)

$=$ $\quad 2 + \frac{8 + 5 + 2}{10}$ (Do the addition.)

$=$ $\quad 2 + \frac{15}{10}$ (There's a common factor of 5 in the fraction.)

$=$ $\quad 2 + \frac{3}{2}$ (The improper fraction will need to be converted to a mixed number.)

$=$ $\quad 2 + 1\frac{1}{2}$ (Now we can do a final addition.)

$=$ $\quad 3\frac{1}{2}$ (At last we have an answer!)

Not every multiplication of two mixed numbers by this method would be quite as lengthy, but you will always get four multiplications and will amount to a big, fiddly calculation.

It is far better to convert the mixed numbers to improper fractions, multiply them out and then, if necessary, convert the answer back to a mixed number. Here is the previous multiplication example done this way:

$$2\frac{1}{2} \times 1\frac{2}{5}$$

$=$ $\quad \frac{5}{2} \times \frac{7}{5}$ (Converted to improper fractions.)

$=$ $\quad \frac{5^1}{2} \times \frac{7}{15}$ (Cancel out a common factor of 5.)

$=$ $\quad \frac{7}{2}$ (The answer is an improper fraction.)

$=$ $\quad 3\frac{1}{2}$ (The answer converted back to a mixed number.)

Isn't that better?

This method also works well if you want to multiply three or more mixed numbers:

$$2\frac{1}{2} \times 1\frac{2}{5} \times 2\frac{1}{8}$$

$= \quad \frac{5}{2} \times \frac{7}{5} \times \frac{17}{8}$ (Converted to improper fractions.)

$= \quad \frac{\cancel{5}1}{2} \times \frac{7}{\cancel{15}} \times \frac{17}{8}$ (Cancel out a common factor of 5, but we still need to multiply 7 by 17.)

$= \quad \frac{119}{16}$ (The improper fraction answer. A long division is needed, not shown here, to get a mixed number.)

$= \quad 7\frac{7}{16}$ (The answer converted back to a mixed number.)

If you attempted to do this example without converting to improper fractions, you would have to do eight separate multiplications and add their results together. The working would be very lengthy and it would be very easy to make a mistake.

Division

First convert to improper fractions. Then it is just like any other rational number division: multiply by the inverse. Here is an example:

$$2\frac{2}{3} \div 1\frac{3}{5}$$

$= \quad \frac{8}{3} \div \frac{8}{5}$ (Converted to improper fractions.)

$= \quad \frac{\cancel{8}1}{3} \times \frac{5}{\cancel{18}}$ (Cancel out a common factor of 8.)

$= \quad \frac{5}{3}$ (The answer is an improper fraction.)

$= \quad 1\frac{2}{3}$ (The answer converted back to a mixed number.)

> **Multiplication and division of mixed numbers should always be done by converting to improper fractions. Then proceed as for any rational number multiplication or division.**

In the examples shown above, the answers turned out to be improper fractions that then had to be converted into a mixed numbers. That doesn't always happen. You might get an integer or a fraction. Here are two examples:

$$3\tfrac{1}{3} \times 1\tfrac{1}{2} \qquad\qquad 1\tfrac{1}{2} \div 4\tfrac{1}{2}$$

$$= \frac{10}{3} \times \frac{3}{2} \qquad\qquad = \frac{3}{2} \div \frac{9}{2}$$

$$= \frac{\cancel{10}5}{\cancel{3}1} \times \frac{\cancel{13}}{\cancel{12}} \qquad\qquad = \frac{3}{2} \times \frac{2}{9}$$

$$= \frac{5}{1} \qquad\qquad = \frac{\cancel{31}}{\cancel{21}} \times \frac{\cancel{12}}{\cancel{39}}$$

$$= 5 \qquad\qquad = \frac{1}{3}$$

Mixed number multiplication can easily create large numerators that then need separate working to get to the answer. Unfortunately, this cannot be avoided. This is one of the reasons why decimals are a better way of doing most mixed number calculations and the next chapter explains how they work.

Practice Exercises

You can download practice examples of the four rules of number with mixed numbers from http://numbersexplained.co.uk.

Chapter 13 – Decimals

Decimals are a convenient way of writing fractions.

As you know, integers are written in a place value system based on 10. We start with units and then, as we add columns to the left, every column is worth 10 of the one on its right:

H   T   U

To show fractions, we add more columns to the right of the units, making each place value worth $\frac{1}{10}$ of the one on its left. A **decimal point** marks where these fraction place values start. A decimal point is either a full-stop (period) or a comma, depending upon which country you are in. I use the full-stop because I am English.

H   T   U   . $\frac{1}{10}$ $\frac{1}{100}$ $\frac{1}{1000}$

To show $\frac{3}{10}$, write a 3 in the $\frac{1}{10}$s place value: 0.3 . We have to use a decimal point to show that the 3 is in the $\frac{1}{10}$s place value – if we just wrote 3 it would read as 3 units. Also, it is usual to write a leading 0 in the units place value because 0.3 is easier to read than just .3 .

Similarly, $\frac{3}{100}$ is 0.03 and $\frac{3}{1000}$ is 0.003.

We don't have to stop at the $\frac{1}{1000}$s place value. For example, $\frac{3}{1\,000\,000}$ is a 3 in the millionths place value which is the sixth column to the right of the decimal point: 0. 000 003. A small gap is written between each group of three for readability.

In theory, you can write very small fractions by adding place values to the right of the decimal point for ever, just as you can write very large numbers by adding place values to the left. In practice, very large and very small numbers become unreadable using decimals. They are better written using a system called standard form which is covered in part 2 of this book.

Decimal fractions are not limited to single digits. For example, $\frac{25}{100}$ is written 0.25 .

You can think of this as $\frac{2}{10}$ and $\frac{5}{100}$ if you wish. This works because $\frac{2}{10} + \frac{5}{100} = \frac{20}{100} + \frac{5}{100}$ = $\frac{25}{100}$, but it is probably better to think of it as 25 hundredths. The place value of the last digit gives the denominator of the fraction.

Similarly, $\frac{125}{1000}$ is 0.125 because it is 125 in the thousandths place value.

How does this work for fractions that do not have $\frac{1}{10}$s, $\frac{1}{100}$s or $\frac{1}{1000}$s as their denominator? For some fractions you can easily change then into fractions that do have suitable denominators. For examples:

$$\frac{1}{2} = \frac{5}{10} = 0.5 \qquad \text{(multiplying numerator and denominator by 5)}$$

$$\frac{1}{4} = \frac{25}{100} = 0.25 \qquad \text{(multiplying numerator and denominator by 25)}$$

$$\frac{1}{8} = \frac{125}{1000} = 0.125 \qquad \text{(multiplying numerator and denominator by 125)}$$

So it turns out that 0.25 and 0.125 are the common fractions $\frac{1}{4}$ and $\frac{1}{8}$.

It is not always obvious how to do this. For example, you might not have spotted that $125 \times 8 = 1000$. Fortunately, there is a procedure that you can always use to convert a fraction into a decimal: divide the numerator by the denominator. This works because the denominator is an instruction to divide.

The following example shows you, step-by-step how you could work out that $\frac{1}{8}$ is 0.125 using this method. The division method used is short division.

$$\begin{array}{r} 0 \\ \hline 1 \end{array}$$ 8 into 1 won't go, so 1 is a remainder

$$\begin{array}{r} 0 \ . \ 1 \\ 8 \, \overline{\smash{)}1 \ . \ {}^{1}0} \end{array}$$ Add a tenths place value containing a 0 and carry the remainder 1. 8 into 10 goes 1 remainder 2.

$$\begin{array}{r} 0 \ . \ 1 \ 2 \\ 8 \, \overline{\smash{)}1 \ . \ {}^{1}0 \ {}^{2}0} \end{array}$$ Add another 0 in the hundredths place value and carry the remainder 2. 8 into 20 goes 2 remainder 4.

$$\begin{array}{r} 0 \ . \ 1 \ 2 \ 5 \\ 8 \, \overline{\smash{)}1 \ . \ {}^{1}0 \ {}^{2}0 \ {}^{4}0} \end{array}$$ Add another 0 in the thousandths place value and carry the remainder 4. 8 into 40 goes 5 times exactly.

You can keep adding zeroes and dividing until you get an answer. Here is another conversion example, $\frac{5}{16}$ shown as a long division:

$$\begin{array}{r} 0 \ . \ 3 \ \ 1 \ \ 2 \ \ 5 \\ 1 \ 6 \, \overline{\smash{)}5^{4} \ . \ {}^{1}0 \ \ 0 \ \ 0 \ \ 0} \\ \underline{4 \quad 8} \\ 2^{1} \ {}^{1}0 \\ \underline{1 \quad 6} \\ {}^{-}4^{3} \ {}^{1}0 \\ 3 \quad 2 \\ 8 \ 0 \\ \underline{8 \ 0} \\ - \end{array}$$

So , $\frac{5}{16}$ is 0.3125 .

There is, however, a snag which appears when you try to work out $\frac{1}{3}$ as a decimal:

$$\begin{array}{r} 0 \ . 3 \ \ 3 \ \ 3 \ \ 3 \\ 3 \, \overline{\smash{)}1 \ . \ {}^{1}0 \ {}^{1}0 \ {}^{1}0 \ {}^{1}0} \end{array}$$ and so on …

Every time you divide 10 by 3, you get a remainder of 1 which gives another 10 in the next place value, so the procedure will go on for ever. This is called a recurring decimal. We write it as $0.\dot{3}$. The dot above the 3 shows that it is the recurring decimal for $\frac{1}{3}$. This distinguishes it from 0.3 which is the terminating decimal for $\frac{3}{10}$. It is true that $\frac{1}{3}$ is quite close to $\frac{3}{10}$ but they are not the same.

The issue of recurring decimals looks like a big problem. In practice it isn't, and here is why. If we use 0.3, which is $\frac{3}{10}$, as a rough value for $\frac{1}{3}$, there is an error of $\frac{1}{3} - \frac{3}{10} = \frac{10-9}{30} = \frac{1}{30}$. This might be good enough for some purposes. If it isn't, we could use 0.33. This is actually $\frac{33}{100}$, and there will be an error of $\frac{1}{3} - \frac{33}{100} = \frac{100-99}{300} = \frac{1}{300}$. That's a lot closer. In fact, we can make the error as small as we like. If, for example, we used 0.3333, the error is now $\frac{1}{3} - \frac{3\,333}{10\,000} = \frac{1\,000 - 9\,999}{30\,000} = \frac{1}{30\,000}$, which is tiny.

Recurring decimals are very common, so there is often going to be a small error in any calculation involving decimals. This doesn't matter, because we can control the amount of error by taking as many decimal places as we need.

When using decimals, you do have to keep this issue of error in mind. This may seem a nuisance, but it is price worth paying because, as you will see in the next chapter, decimal fraction calculations are so much easier than working with fractions. (Error is examined in more detail in Part 2 of this book.)

A real number is a sequence of digits with a decimal point somewhere. (There are more rigorous definitions.) The set of real numbers is infinitely large.

All rational numbers can be converted into real numbers and they form either a sequence of digits that terminates or an infinitely long sequence that recurs.

There are also real numbers that are irrational – they consist of an infinitely long sequence of digits that do <u>not</u> recur. The best known example of this is the number π (pi), which is used in calculations involving circles and things that oscillate.

There is, in fact an infinitely large set of irrational numbers. Irrationals are the subject of a chapter in part 2.

Real numbers are continuous – no matter how close two real numbers are, you can always find more real numbers in between.

They are also uncountable. This means that they cannot be put into one-to-one correspondence with the set of positive integers. How weird is that?

More About Recurring Decimals

It is very common for a fraction to convert to a recurring sequence of digits. Here is an example, $\frac{4}{11}$:

$$\begin{array}{r} 0\ .\ 3\ \ 6\ \ 3\ \ 6 \\ \hline 1\ 1\ \overline{\big)\ 4\ .\ ^40\ ^70\ ^40\ ^70} \end{array} \quad \text{and so on ...}$$

We write this as $0.\dot{3}\dot{6}$. A dot goes above the beginning and end of the sequence. There is a subtly different recurring decimal for $\frac{7}{11}$:

$$\begin{array}{r} 0\ .\ 6\ \ 3\ \ 6\ \ 3 \\ \hline 1\ 1\ \overline{\big)\ 7\ .\ ^70\ ^40\ ^70\ ^40} \end{array} \quad \text{and so on ...}$$

So, $\frac{7}{11}$ is $0.\dot{6}\dot{3}$. This is not the same as $\frac{4}{11}$ which is $0.\dot{3}\dot{6}$. It is the same recurring sequence but starting at a different digit – a small but significant difference. (You might be interested to try working out the decimal equivalents of the other elevenths fractions: $\frac{1}{11}$, $\frac{2}{11}$ and so on. What do think you might find?)

The recurring sequence can be quite long. Here, for example, is $\frac{1}{7}$:

$$\begin{array}{r} 0\ .\ 1\ \ 4\ \ 2\ \ 8\ \ 5\ \ 7\ \ 1 \\ \hline 7\ \overline{\big)\ 1\ .\ ^10\ ^30\ ^20\ ^60\ ^40\ ^50\ ^10} \end{array} \quad \begin{array}{l}\text{and now the remainder 1 recurs,}\\ \text{and so does the answer}\end{array}$$

So, $\frac{1}{7}$ is $\dot{1}4285\dot{7}$.

Notice that the decimal for $\frac{1}{7}$ has a six digit recurring sequence. This is the maximum length, because the remainders when dividing by 7 can only be 1, 2, 3, 4 , 5 and 6. Once all the possible remainders have been used, the sequence must recur. However, not all recurring decimals use this maximum number of digits – you have already seen that elevenths only use 2 digits.

The question of how long the recurring sequence will be for a particular divisor is another problem in mathematics that currently has no solution.

You might like to try working out the decimal equivalents of the other sevenths fractions: $\frac{2}{7}, \frac{3}{7}$ and so on. What do you notice? Do you need to divide them all out completely? Go on, give it a try and see how quickly you can spot what is going on. You might also find the ninths interesting.

Sometimes a decimal has a short non-recurring sequence before it recurs. Here is an example, $\frac{5}{36}$:

$$
\begin{array}{r}
0\ .\ 1\ \ 3\ \ 8\ \ 8 \\
36\overline{\smash{\big)}\,5^4\ .\ {}^10\ \ 0\ \ 0\ \ 0} \quad \text{and so on ...} \\
\underline{3\quad 6} \\
1\quad 4^3\ {}^10 \\
\underline{1\quad 0\quad 8} \\
3^2\ {}^1\!2^1\ {}^10 \\
\underline{2\quad 8\quad 8} \\
3^2\ {}^1\!2^1\ {}^10 \\
\underline{2\quad 8\quad 8} \\
3\quad 2
\end{array}
$$

So, $\frac{5}{36} = 0.13\dot{8}$ where there are two non-recurring digits 13 before a recurring 8.

Converting Fractions to Decimals in Summary

Sometimes it is obvious, for example $\frac{3}{10} = 0.3$.

You can often get a decimal equivalent by looking at a related fraction. For instance, if $\frac{1}{4}$ is 0.25, then $\frac{3}{4}$ must be 3 times that: 0.75. Also, $\frac{1}{40}$ must be $\frac{1}{10}$ of 0.25 which just means moving everything one place to the right: 0.025 . It is easier to think of this as moving the decimal point one place to the left.

You can sometimes see immediately what the decimal equivalent of a fraction is, or you can see an easy way to work it out, but if you can't, this is what you do:

> **To change from a fraction to a decimal, divide the numerator by the denominator, adding as many zero decimal places as you need, until the answer terminates or recurs.**

Numbers Explained

Converting Decimals to Fractions

Converting decimals back into fractions is something that you do not often need to do, but it is not difficult. Here's what you do:

> **To change a decimal into a fraction, take the place value of the last digit and make this the denominator. Then write the complete sequence of digits, without the decimal point or any leading zeroes, as the numerator.**

For example, 0.19 is $\frac{19}{100}$. This is in its lowest terms because 19 is a prime number.

For another example, 0.875 will be $\frac{875}{1000}$. Then reduce this to its lowest terms, which is most easily done in stages, dividing out common factors of 5:

$$= \frac{875}{1000} = \frac{175}{200} = \frac{35}{40} = \frac{7}{8}$$

0.875 is $\frac{7}{8}$.

Converting Recurring Decimals

The easiest thing to do is to choose how accurate you want to be and get a fraction that is close to the true answer. For example, $0.\dot{4}\dot{3}$ is roughly $\frac{43}{100}$. A better answer would be $\frac{434}{1000}$ which is $\frac{217}{500}$ in its lowest terms. And so on, to get fractions which are closer and closer to the true value.

If you want to find the true fraction equivalent of a recurring decimal, a little algebra is involved. The ability to do this is not important so you may want to skip this procedure, although it is an elegant bit of maths and not too difficult to understand.

Suppose we want to know the fraction equivalent of $0.\dot{6}$. (You may already know that this is actually the decimal for $\frac{2}{3}$ - not surprisingly because it is $2 \times \frac{1}{3}$. Since $\frac{1}{3}$ is $0.\dot{3}$, it follows that $\frac{2}{3} = 0.\dot{6}$, with all the decimals digits multiplied by 2.)

Here's the procedure. We can use the variable x to represent the answer:

$$x = 0.\dot{6}$$

Then it follows that $10x = 6.\dot{6}$ (shifting the decimal point 1 place to the right)

Subtracting x from $10x$ gives $9x$, so it must be true that:

$$9x = 10x - 9x = 6.\dot{6} - 0.\dot{6}$$

All of the decimal digits in this subtraction will give zero and be dropped, so

$$9x = 6$$

Dividing both sides of the equation by 9 gives:

$$x = \frac{6}{9} = \frac{2}{3} \text{ which is the accurate answer.}$$

For longer recurring sequences, we use the same trick. For instance, earlier we got rough fraction equivalents for $0.\dot{4}\dot{3}$. Here is the true equivalent fraction:

Let $x = 0.\dot{4}\dot{3}$, then $100x = 43.\dot{4}\dot{3}$

Subtracting x from $100x$, gives $99x$, so we get:

$$99x = 43.\dot{4}\dot{3} - 0.\dot{4}\dot{3}$$

So, $99x = 43$

Dividing both side of the equation by 99 gives:

$$x = \frac{43}{99} \text{ which is in its lowest terms because 43 is a prime.}$$

If it were a 3 digit recurring sequence, you would do the same procedure but with $1000x$. A 4 digit sequence would require $10\,000x$, and so on.

Elegant, isn't it?

Practice Exercises

You can download practice examples of converting fractions into decimals, and decimals to fractions, from http://numbersexplained.co.uk.

Chapter 14 – The 'Four Rules' for Decimals

This chapter shows you how to add, subtract, multiply and divide decimal fractions. Doing calculations with decimals is much less complicated than working with the mixed numbers that they represent.

Addition and Subtraction

All that you need do is make sure that the place values are lined up correctly and then just add and subtract as if the numbers were integers. The decimal points should be all in line, and that is where the decimal point will be in the answer. You might have to put in a few zeroes after the decimal point to make a subtraction work, but that is as hard as it gets. Here are some examples:

24.8 + 3.99 :

$$\begin{array}{r} 2\,4\,.\,8 \\ 3\,.\,9\,9 \\ \hline 2\,8\,.\,7\,9 \\ \tiny 1 \end{array}$$

(If there is nothing in a place value, treat it as a zero.)

The answer is 28.79

0.04 + 0.007 :

$$\begin{array}{r} 0\,.\,0\,4 \\ 0\,.\,0\,0\,7 \\ \hline 0\,.\,0\,4\,7 \end{array}$$

(You could do this in your head.)

The answer is 0.047

1.08 - 0.107 :

$$\begin{array}{r} \cancel{1}^{0}\,.\,{}^{1}0\,\cancel{8}^{7}\,{}^{1}0 \\ 0\,.\,1\,\ 0\,\ 7 \\ \hline 0\,.\,9\,\ 7\,\ 3 \end{array}$$

(You have to put in a zero in the $\frac{1}{1000}$s place value to make the subtraction work.)

The answer is 0.973

4 - 0.01 :

$$\begin{array}{r} 4^{3}\,.\,{}^{4}\cancel{0}^{9}\,{}^{1}0 \\ 0\,.\,0\,\ 1 \\ \hline 3\,.\,9\,\ 9 \end{array}$$

(You could do this in your head by counting backwards by 0.01 from 4.)

The answer is 3.99

> **To add and subtract decimals, line up the decimal points.**

Shifting the Decimal Point

If you shift the decimal point in a number, you are multiplying or dividing by 10 or 100, or 1000, etc. This is a common and useful type of calculation.

For example, if you take a decimal such as 2.35 and multiply it by 10, then the 2 will become 20 and the 3 tenths will become 3 units and the 5 hundredths will become 5 tenths. This means that all three digits will move one place value to the <u>left</u>, like this:

   Which is the same as moving the decimal point one place to the <u>right</u>

If you multiply by a hundred, the 2 will become 200 and the 3 tenths will become 30 and the 5 hundredths will become 5 units. This means that all three digits will move <u>two</u> place values to the left and the decimal places will disappear, like this:

   Which is the same as moving the decimal point <u>two</u> places to the right

If you multiply by a thousand, the 2 will become 2000 and the 3 tenths will become 300 and the 5 hundredths will become 50. This means that all three digits will move <u>three</u> place values to the left, the decimal places will disappear and you will need to put a zero in the empty units place value, like this:

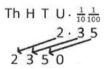   Which is the same as moving the decimal point <u>three</u> places to the right

It is easier to think of it in terms of moving the decimal point than all the digits changing columns. When you multiply by 10 or 100 etc., here's how it works:

1. The decimal point moves right by the same number of places as there are zeroes in the number you are multiplying by e.g. 3 places for x 1000.
2. Drop any leading zeroes if they are no longer needed.
3. Add trailing zeroes in the units, hundreds and higher place values as necessary.
4. If there are no longer any decimal places, drop the decimal point.

These rules will become clear if you study these examples:

$0.002 \times 10 = 0.02$ — Move the decimal point one place to the right and drop a leading zero.

$0.025 \times 100 = 2.5$ — Move the decimal point two places to the right and drop both leading zeroes.

$5.7 \times 1000 = 5\,700$ — Move the decimal point three places to the right. This means you need two trailing zeroes. Drop the decimal point.

$1.65 \times 1\,000\,000 = 1\,650\,000$ — Move the decimal point six places to the right and drop it. Four trailing zeroes are needed.

$0.03 \times 1\,000 = 30$ — Move the decimal point three places to the right. Drop the leading zeroes and one trailing zero is needed.

$314 \times 100 = 31\,400$ — Move the decimal point, which is unwritten, two places to the right. Two trailing zeroes are needed.

In the last example, this is the 'add zeroes' procedure for multiplying integers which is a special case of the more general procedure for decimals that you have been shown here.

For dividing by 10, 100, 1000 etc., the movement is in the opposite direction. If, for example, you divide the number 42.7 by 10, the 4 in the tens column will become 4 units, the 2 units will become 2 tenths and the 3 tenths will become 3 hundredths. This means that all three digits will move one place value to the <u>right</u>, like this:

H T U · $\frac{1}{10}\frac{1}{100}$
4 2 · 7
4 · 2 7

Which is the same as moving the decimal point one place to the <u>left</u>

4 · 2 7
Now Was
here here

Again, it is easier to think of it in terms of moving the decimal point than all the digits changing columns. So, when you divide by 10 or 100 etc. :

1. If there is no decimal point because you are starting with an integer, put a decimal point just after the units and move it from there.
2. The decimal point moves left by the same number of places as there are zeroes in the number you are dividing by e.g. 3 places for ÷ 1000.
3. Add leading zeroes as necessary.
4. Drop any trailing zeroes if they are no longer needed.

Here are some examples:

$138.5 \div 10 = 13.85$

Move the decimal point one place to the left.

$0.02 \div 10 = 0.002$

Move the decimal point one place to the left and add a leading zero.

$3.58 \div 100 = 0.0358$

Move the decimal point two places to the left. This will require two leading zeroes.

$82.9 \div 1\,000\,000 = 0.000\,082\,9$

Move the decimal point six places to the left. This will require five leading zeroes.

$$115 \div 1000 = 0.115$$

Put a decimal point just after the units and move it three places to the left. A leading zero will be needed.

$$2500 \div 1000 = 2.5$$

Put a decimal point just after the units and move it three places to the left. Drop the two trailing zeroes.

Multiplying or dividing by 10, 100, 1000, etc. involves moving the decimal point by the number of zeroes in the divisor.

Move the decimal point to the right to multiply.

Move the decimal point to the left to divide.

Multiplication of Decimals

Consider an example of two decimal numbers being multiplied together:

$$0.2 \times 0.03$$

If we write this in the form of fractions, it looks like this:

$$\frac{2}{10} \times \frac{3}{100}$$

Without reducing to lowest terms, the answer to this is:

$$\frac{2}{10} \times \frac{3}{100} = \frac{6}{1000}$$

Writing this as decimals, we get:

$$0.2 \times 0.03 = 0.006$$

Notice that there is one decimal place in the first number, two decimal places in the second, and there are three decimal places in the answer. If you add the number of decimal places in the question, you get the number of places in the answer.

Similarly:

$$0.02 \times 0.003 = 0.000\,06 \qquad \text{(In decimal places, 2 + 3 = 5)}$$

because, in fractions, this is

$$\frac{2}{100} \times \frac{3}{1000} = \frac{6}{100\,000}$$

This works for larger numbers. For example:

$$2.4 \times 3.9 = 9.36 \qquad \text{(In decimal places, 1 + 1 = 2)}$$

because, in fractions, this is

$$\frac{24}{10} \times \frac{39}{10} = \frac{936}{100}$$

Working

	2	4
	3	9
2	1^3	6
7^1	2	0
9	3	6

When you multiply decimals, ignore the decimal points and multiply as if the numbers were integers. When you have the answer, count up how many decimal places there are altogether in the question and this will be the number of decimal places in the answer.

There is one thing to watch out for. If one of the numbers is an integer, do not ignore any zeroes when counting decimal places. For example suppose you multiply 35 000 by 0.25.

First you need to calculate 35 x 25. The answer is 875.

If 35 x 25 is 875, then 35 000 x 25 is 1000 times larger: 875 000.

Now adjust the decimal point. There are 2 decimal places in the question, so two of those zeroes will be decimal places and they can be dropped as you move the decimal point two places left: 875 0.~~00~~

Working

	3	5
	2	5
1	7^2	5
7^1	0	0
8	7	5

The answer is 8750.

If you didn't take the zeroes into account, you would get 8.75 which would be 1000 times too small.

Numbers Explained

Division

The procedure for dividing decimals depends upon the fact that you will get the same answer if you divide 4 by 2 or if you divide 40 by 20. The answer to both divisions is 2.

We can go further. There are an infinite number of similar divisions based on $4 \div 2 = 2$. Here are some of them. (For convenience I have used the notation of rational numbers e.g. $\frac{4}{2}$ to show the division $4 \div 2$.)

$$\frac{4}{2} = \frac{40}{20} = \frac{400}{200} = \frac{4000}{2000} = \frac{40\,000}{20\,000} \text{ and so on} = \frac{2}{1} = 2$$

In each case the numerators and denominators have been multiplied by 10 or 100 or 1000 etc.. This causes the decimal point in each number to shift to the right.

Similarly, we can divide the numerators and denominators by 10 or 100 or 1000 etc.. This will shift the decimal points to the left, like this:

$$\frac{4}{2} = \frac{0.4}{0.2} = \frac{0.04}{0.02} = \frac{0.004}{0.002} = \frac{0.0004}{0.0002} \text{ and so on} = \frac{2}{1} = 2$$

This will work for any division, not just $4 \div 2$. We can move the decimal points to the left or right by the same number of places in both numbers of a division without changing the answer. For example, $1.5 \div 0.3 = 5$ is the same as $15 \div 3$, moving the decimal points one place to the right. This is how we divide a number by 0.3 – move the decimal points one place to the right and divide by 3. Similarly, if you want to divide by 0.7, move the decimal points one place to the right and divide by 7.

Moving the decimal points is also helpful in dividing large integer numbers. For example, $24\,000 \div 80$ is more easily calculated as $2\,400 \div 8$ giving the answer 300 .

So, using this principle, here is how you divide decimals. Move the decimal point in the divisor so that its last digit is in the units place value. Move the decimal point the same number of places in the number you are dividing. Then do the division.

Here are some examples, all using the division $48 \div 4 = 12$ so that you can concentrate on examining the shifts in place value:

$480 \div 0.4 = 4800 \div 4 = 1200$ (move 1 place right)

$0.048 \div 0.4 = 0.48 \div 4 = 0.12$ (move 1 place right)

$4.8 \div 0.04 = 480 \div 4 = 120$ (move 2 places right)

$480 \div 400 = 4.8 \div 4 = 1.2$ (move 2 places left)

$0.0048 \div 0.004 = 4.8 \div 4 = 1.2$ (move 3 places right)

$480 \div 0.000\,4 = 4\,800\,000 \div 4 = 1\,200\,000$ (move 4 places right)

If the actual division requires some working, make sure that the decimal point in the answer is lined up with the decimal point in the question. For example:

$$10.24 \div 0.8 = 102.4 \div 8 = 12.8$$

$$
\begin{array}{r}
1\ 2\,.\,8 \\
\hline
8\,|\,1\ {}^{1}0\ {}^{2}2\,.\,{}^{6}4
\end{array}
$$

$$38.7 \div 600 = 0.387 \div 6 = 0.0645$$

$$
\begin{array}{r}
0\,.\,0\ 6\ 4\ 5 \\
\hline
6\,|\,0\,.\,3\ {}^{3}8\ {}^{2}7\ {}^{3}0
\end{array}
$$

Here is a final example requiring a long division, but the principle is the same: adjust the decimal point and then divide. (The working calculations of the multiples of 125 are not shown.)

$$54.75 \div 12.5 = 547.5 \div 125 = 4.38$$

(Note that another zero is required in the hundredths place value before the division terminates. In general, you just keep dividing until you have an answer that is sufficiently accurate.)

$$
\begin{array}{r}
4.38 \\
\hline
125\,|\,547.50 \\
500 \\
\hline
47.5 \\
37.5 \\
\hline
10.00 \\
10.00 \\
\hline
0
\end{array}
$$

To divide decimals, move the decimal point in the divisor so that its last digit is in the units place value. Move the decimal point the same number of places in the number you are dividing. Do the division. The decimal point in the answer will line up with the adjusted decimal point in the question.

Avoiding Long Divisions

Long division is a bit of a pain. It has to be carefully laid out and you will often have to do additional bits of working in order to get the multiples of some long number. If you can avoid it, do so.

Sometimes you can avoid it by factorising the divisor. The last example of decimal division required the division $547.5 \div 125$. But 125 is 5×25. In fact, it can be broken down further into $5 \times 5 \times 5$. This means that $547.5 \div 125$ can be broken down into $(((547.5 \div 5) \div 5) \div 5)$. This looks complicated with its nested brackets, but what it says is "divide 547.5 by 5, divide the answer by 5 again, and then divide by 5 a third time".

The best way to lay this calculation out is as a stack of short divisions like this:

$$
\begin{array}{r}
1\ 0\ 9\ .\ 5 \\
\hline
5\,|\,5\ 4\ ^4 7\ .\ ^2 5
\end{array}
$$

Now divide the answer by 5 again

$$
\begin{array}{r}
2\ 1\ .\ 9 \\
\hline
5\,|\,1\ ^1 0\ 9\ .\ ^4 5 \\
\hline
5\,|\,5\ 4\ ^4 7\ .\ ^2 5
\end{array}
$$

Finally, divide by 5 a third time.

$$
\begin{array}{r}
4\ .\ 3\ 8 \\
\hline
5\,|\,2\ ^2 1\ .\ ^1 9\ ^4 0 \\
\hline
5\,|\,1\ ^1 0\ 9\ .\ ^4 5 \\
\hline
5\,|\,5\ 4\ ^4 7\ .\ ^2 5
\end{array}
$$

The answer is 4.38

Always look to see if you can factorise the divisor like this and replace a difficult long division with two or more short divisions. Here are couple more examples:

$7.644 \div 9.8$.

Factorise like this: 98 is even, so it is 2×49 and 49 is 7×7. So, shift the decimal point 1 place right, divide by 2 and then divide twice by 7 :

```
        0 . 7  8
    7 | 5 . ⁵4 ⁵6
    7 | 3 8 . ³2 ⁴2
    2 | 7 ¹6 . 4  4
```

The answer is 0.78

Convert $\dfrac{1}{72}$ to a decimal.

You can avoid dividing 1 by 72 if you factorise 72 into, say, 8×9 . (There are other possibilities such as $3 \times 4 \times 6$ — they would all work.) Here we will divide by 8 first because it will terminate and not recur. Then we will divide the answer by 9 until we can see the recurring sequence:

```
      0 . 0  1  3  8  8
  9 | 0 . 1 ¹2 ³5 ⁸0 ⁸0
  8 | 1 . ¹0 ²0 ⁴0
```

The answer is $0.013\dot{8}$.

Practice Exercises

You can download practice examples of the four rules with decimal numbers from http://numbersexplained.co.uk.

Numbers Explained

Chapter 15 – Measuring, Accuracy and Estimation

Decimals give us a convenient way of representing rational numbers and give us much easier ways of working. The problem with decimals is that quite common fractions turn out to be recurring sequences of digits that go on forever and we have to use shortened versions of them. This means that working with decimals will often give us slightly inaccurate answers.

In practice this is not a problem. In the real world we are often dealing with measurements that have only limited accuracy anyway.

This chapter looks in more detail at accuracy and then goes on to the related and important topic of estimation. First, however let's consider measurements.

Measurements

The labels that we attach to measurements are called **units**. The basic units of measurement are those of time, length and mass. (If you want to know more about this and about other units such as electrical measurements, I suggest you have a look in Wikipedia. This is a huge topic and we will only touch on the essentials here.) We will also look at the units of capacity (volume).

Time

The basic unit of time is the **second**. Originally, a second was a fraction of a day. 60 seconds = 1 minute and 60 minutes = 1 hour. Then 24 hours = 1 day, so 1 second is $\frac{1}{60 \times 60 \times 24} = \frac{1}{86400}$ of a day. Then 7 days = 1 week. After that it gets messy, with a variable number of days to a month and a year because these depend upon the time it takes for the earth to spin on its axis (the day), the moon to orbit the earth (the lunar month) and the earth to orbit the sun (the solar year). The solar day (noon to noon) varies by up to 30 seconds depending on the season. The lunar month (full moon to full moon) is 29 days, 12 hours, 44 minutes and 3 seconds, so we vary the number of whole days in a month with one month being 28 or 29 days and the rest either 30 or 31 days. The solar year is 365.242 days, which is close to 365.25, so we can use 365 for three years followed by a 366 day year and it doesn't go too far wrong. All these statements are simplifications of some very complicated issues.

Why 60? It is thought that this goes back to at least 2000 BC when the ancient civilisations of the Middle East used base 60 numbers. An advantage of using 60 instead of a decimal number like 10 or 100 is that you get an exact value for thirds and quarters – 20 seconds and 15 seconds.

This is probably why we use base sixty units for time and also for sub-units of angles. The most common measurement unit for small amounts of turning, or angles, is the degree. (Strictly, it is a degree of arc, to distinguish it from a degree of temperature.) A sixtieth of a degree is a minute of arc and a sixtieth of a minute is a second or arc.

In 1960, the modern SI (standing for Système Internationale) abandoned astronomical definitions and started using properties of atoms to define the second because these are more regular and reliable.

The second is supposed to be abbreviated to s, but is often shown as sec.

Smaller units are the millisecond (abbreviated to ms), which is a thousandth of a second, and the microsecond (μs) which is a millionth of a second.

SI Prefixes

Notice how the prefixes milli- and micro- were attached to the second. Modern measuring units (often called metric units) all employ these prefixes to create smaller and larger units from the basic unit. The commonly used prefixes for small units are:

Prefix	Abbreviation	Meaning	
deci-	d	Tenth	
centi-	c	Hundredth	
milli-	m	Thousandth	
micro-	μ	Millionth	(μ is the Greek letter Mu.)

This is not a complete list, just the commonly used ones.

The commonly used prefixes for large units are:

Prefix	Abbreviation	Meaning
kilo-	k	Thousand
mega-	M	Million (note the upper-case abbreviation)

Length

There are dozens of old fashioned units for length. Most countries, apart from the USA, have largely abandoned oddly defined units such as 12 inches to the foot, 3 feet to a yard, 22 feet to the chain (a surveyors chain – also the length of a cricket pitch) – 10 chains to the furlong (a furrow length from medieval land measurement but now only used in horse-racing) and 8 furlongs to the mile (which makes 1 mile the same as 1760 yards). It is easy to see why they are not used so much now – they are a nightmare when you start making calculations with them. The mile is still widely used, and there is also the nautical mile. A nautical mile is much longer at 1,852 metres (about 2025 yards) and is given this length because it is 1 minute of latitude.

The modern SI unit of length is the **metre**, invented in early nineteenth-century France and once defined as the length of a bar of platinum in Paris. It now has a definition based on the properties of an atom of krypton. It is abbreviated to m.

Common units are:

Unit	Abbreviation	Meaning	Comments
kilometre	km	1000m	Approximately $\frac{5}{8}$ mile
centimetre	cm	0.01 metre	Replaces the inch. 2.54cm = 1 inch
millimetre	mm	0.001m	Building measurements are usually given in mm because you can work to the nearest mm for most purposes.
micrometre	μm	0.000 001m	Also called the micron

Mass

Mass is often confused with weight and force and I do not want to complicate this book by explaining the distinctions, which are not important for everyday purposes.

As with length, there are lots of ancient crazy systems of units such as avoirdupois, which goes 16 ounces (oz) = 1 pound (lb), 14 lb = 1 stone, 2 stone = 1 quarter, 4 quarters = 1 hundredweight (cwt), 20 cwt = 1 ton, which makes the ton 2240 lb. Notably, the pound is still in use in the USA and tons are commonly used everywhere, but most countries have gone over to metric units.

The basic unit used to be the gramme, now usually written gram, but this is inconveniently small, so it is now the **kilogram**. The metric units of mass or weight are:

Unit	Abbreviation	Meaning	Comments
gram	g		500 gm is often called a pound. The actual value is 1 lb ≈ 454 g
kilogram	kg	1000 g	
tonne	t	1000 kg	Also called the metric ton because, by happy coincidence, 1 tonne is roughly the same as 1 ton

Capacity

Pints and gallons (and pecks and bushels and barrels) have now been mostly superseded by the metric unit, the **litre**. The abbreviation is l.

Capacity, or volume, can always be expressed in terms of a cube of given length and this is actually the definition of the litre: 1l = 1000 cubic cm.

The cubic centimetre (often abbreviated to cc) was a commonly used unit until recently but is now replaced by its exact equivalent the millilitre (ml).

Large volumes are usually given in cubic metres and it is worth remembering that a cubic metre is 1000 l.

It is also worth remembering that 1l of water weighs 1 kg. (This is actually how the kilogram is defined. The definition is under review because it is not a fundamental property of nature that can be readily reproduced in a laboratory.) Another useful fact is that a cubic metre of water weighs 1 tonne or approximately 1 ton.

We tend to use litres to measure fluid quantities and use cubic measures for things that have a fixed shape.

Numbers Explained

Counting Versus Measuring

Counting and measuring are often confused.

Counting answers the question "How many?" and the answer will be an integer. For example, you might ask "How many passengers can that car carry?". The answer might be "4 adults." Note that 'adults' is in the plural and the answer tells us the number of passengers. Comparisons use the words 'more' and 'fewer'.

Measuring answers the question "How much?" and the answer will be a continuous number such as a fraction, mixed number or a decimal. For instance, you might ask "How much fuel is in the tank?" The answer might be "10.22 litres of diesel." Note that the word 'diesel' is in the singular and the answer is an amount. Comparisons use the words 'more' and 'less'.

For some reason, it has become commonplace for people to use the word 'amount' when they mean 'number'. It does not make any more sense to talk about an 'amount of passengers' in the car than it would to talk about a 'number of diesel'. It is also incorrect to talk about a car that can carry 'less' passengers – it should be 'fewer'.

Accuracy

This section is concerned with the measurement and the accuracy of calculations based on measurements. Accuracy is often called 'degree of accuracy'. There are two ways of specifying accuracy: decimal places and significant figures.

Decimal Places

If we are only working with numbers less than about 100, then it is good enough to say how many decimal places we are going to use. 'Decimal places' is usually shortened to 'dp'. For instance, we might have a number given as:

    2.42 to 2 dp   or   2.42 (2 dp)

What this means is that we are counting in hundredths. It tells us that the nearest number to the true figure using this scale is 2.42. It can be helpful to consider what numbers are <u>not</u> as close. Counting in hundredths, we are saying that the nearest value is not 2.41 or 2.43.

The following diagram shows the range of values that are closest to 2.42:

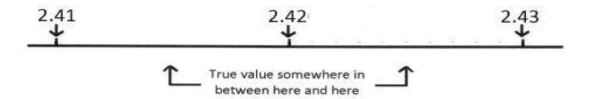

If the number were shown as 2.420 to 3 dp, this would mean that we are counting in thousandths – the nearest value to the true figure is 2.420 and not 2.419 or 2.421. This diagram shows the much narrower range of values that are closest to 2.420:

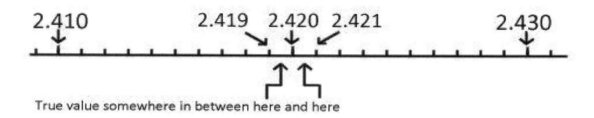

What about 4dp? If the number was given as 2.4200 to 4dp, this means that the nearest value to the true figure is 2.4200 and not 2.4199 or 2.4201. You will have to imagine this diagram because the numbers are so close together that it can't be drawn to the same scale as the previous two.

These comparisons show why we can safely ignore most of a recurring sequence of decimals, as long as we are aware of what we are doing. As the number of decimal places increases, any error between the true value and the value we are using becomes so small that it doesn't matter.

Significant Figures

The more common way of specifying accuracy is to state how many significant figures there are. Every place value is one tenth of the one on its left. So as you read the digits of a number from left to right, they become steadily less significant until you decide that they can be ignored. Three significant figure accuracy, written '3sf', means that 3 digits are good enough. However, there may be zeroes in the number that are there to indicate the place value. The following numbers have 3sf accuracy and the significant digits are in bold so that you can see which zeroes are significant and which are there only to fill up empty place values:

2.04 (3sf)   0.00**2 04** (3sf)   **20 4**00 (3sf)

Here is the same number but with 4 significant figure accuracy:

2.040 (4sf)   0.00**2 040** (4sf)   **20 40**0 (4sf)

How Accurate Can You Be?

Three or four significant figure accuracy is usually good enough for most purposes. Sometimes 5sf is needed, but consider how accurate this is. If you were measuring a distance of about 4 metres and you gave a measurement as 4.2308 m, you are measuring to the nearest tenth of a millimetre. You are claiming that the distance is not 4.2307 or 4.2309. Really? A tenth of a millimetre is hard to see with the unaided eye. Also, you would need special instruments to perform the measurement – a tape measure wouldn't do.

Beware of spurious accuracy. If you are using a calculator, it knows nothing about accuracy and will often fill up its display with long strings of digits. Most of them are meaningless and you should ignore them. This is called rounding an answer.

Rounding

Suppose that you have a number that is to be rounded to 3 sf. All you need do is look at the 4th significant figure. If it is:

- between 0 and 4, round down by leaving the 3rd significant figure as it is
- between 5 and 9, round up by adding 1 to the 3rd significant figure

Then, discard any digits after the 3rd significant figure.

Here are some examples of rounding to 3 sf:

4.342	the 4th sf is a 2 so round down, giving:	4.34
4.347	the 4th sf is a 7 so round up, giving:	4.35
0.0055682	the 4th sf is an 8 so round up, giving:	0.00557
33 333.$\dot{3}$	the 4th sf is a 3 so round down, giving	33 300 (3sf)
2.397	the 4th sf is a 7 so round up, giving:	2.40

In the last two examples, the trailing zeroes are there for different reasons:

- In 33 300, the two zeroes are just there to show the place value (the number isn't 33 200 or 33 400). However, you cannot know this unless you write 3sf after the number to show the accuracy.
- In the 2.40, the zero is significant (it isn't 2.39 or 2.41 and you wouldn't write it if it were not significant. The 3sf accuracy is implied by the zero.

Rounding up can give some surprising results when there are sequences of 9s. For instance, if you round 49.9733 to 3sf, you will get 50.0 . This is because the 4th sf is a 7, which means round up, and when you add 0.1 to 49.9, you get 50.0 .

You can round to other degrees of accuracy than 3sf. Here are some examples:

4.342	rounded to 1dp:	4.3
53.75	rounded to 2sf:	54
18 456 221	rounded to 4sf:	18 460 000
18 456 221	rounded to 1sf:	20 000 000
0.01$\dot{6}$	rounded to 2dp:	0.02
0.0008	rounded to 2sf	0.00

Look carefully at the last example: 0.00 is not the same as 0. The first zero is not significant and is just there to emphasise the decimal point. The two zeroes after the decimal point are significant and they tell you that the number is closer to 0.00 than it is to 0.01 . The number is small but it isn't necessarily 0.

Do not fall into the trap of rounding up one digit at a time. If you were to round 0.32747 by first rounding to 4sf, giving 0.3275, and then rounding again to 3sf, giving 0.328, you will have rounded the wrong way. Why? It's because 0.32747 is closer to 0.327 (the difference is 0.00047) than it is to 0.328 (the difference is 0.00053) . Always look at the next digit (it's a 4, so round down) and you won't go wrong.

When you do a calculation, good practice is to work to one or more degree of accuracy than you need. For example, you might decide to work with at least 4sf accuracy. Then, when you have the answer, round it to 3sf. A useful symbol is the 'approximately equal' sign, which is usually an equals sign with a wobbly top line: $\cong$. (There are several variations on this sign such as $\approx$, but the meaning should be obvious from the context.)

Here is an example: 4.567×23.15 . If you do this on a calculator you will get 105.72605 . The calculator doesn't know any better so it has given you an answer with 8sf . Most of the decimal places in this answer are nonsense because the answer can't be more accurate than the numbers you start with. We started with 4sf accuracy, so round the answer to 3sf : 106 .

It is the least accurate number in a calculation that will determine the accuracy of the answer. For instance, you might want to calculate 0.2 x 3.553. The product of the two numbers is 0.7106. Although 3.553 is accurate to 4sf, 0.2 is only 1sf so you should only claim 1sf accuracy, which is 0.7 .

Estimation

Estimation is looking at a calculation and working out a rough answer in your head. It doesn't matter whether you do this before or after you work out an accurate answer, but it is important that you always do it so that you can check for gross mistakes in the calculation.

It is particularly important to estimate answers if you are using a calculator. It is true that calculators don't make mistakes, but it is very easy to have a bit of finger trouble when entering the calculation. This also true of spreadsheets.

There is no single procedure for performing estimates. It depends upon the nature of the calculation. It can also depend upon the actual numbers involved because you can sometimes spot ways of simplifying a calculation. The more familiar you become with number work, the easier this gets. Here are some of the things that you can do.

Estimating Multiplications and Divisions

A routine way of estimating the answer to a calculation involving multiplication or division is to round everything to 1sf and work it out. Here are some examples:

Calculation	Rounded Calculation	Estimate	Actual Answer (by calculator)
4.7×3.2	5×3	15	15.04
$87.3 \div 2.4$	$100 \div 2$	50	36.375
350×450	300×500	150 000	157 500
$7.5 \div 0.667$	$\begin{aligned} & 7 \div 0.7 \\ = \ & 70 \div 7 \end{aligned}$	10	11.244
$\dfrac{49.97 \times 12.8}{0.44}$	$\begin{aligned} & \dfrac{50 \times 10}{\frac{1}{2}} \\ = \ & 500 \times 2 \end{aligned}$	1000	1453.7

In the last example, the 0.44 hasn't been rounded to 0.4. Instead it was rounded up to 0.5 and changed into the fraction $\frac{1}{2}$. This is because dividing by half is the same as multiplying by 2 which is easier than dividing by 0.4. This is the sort of thing that you can do to make an estimate easy.

Switching to easy fractions is often helpful with divisions. For example, if you have a calculation with $\div\, 0.085$, rounding to $\div\, 0.09$ isn't very helpful. But if you round it a bit further to $\div\, 0.1$, that is the same as $\div\, \frac{1}{10}$ which is the same as $\times 10$.

It is also useful to consider if you are expecting the answer to be more or less than your estimate. If you have rounded both numbers down in a multiplication, then you would expect your estimate to be less than the true answer. The opposite would be true if both numbers were rounded up.

Remember that numbers get bigger if you multiply by a number bigger than 1 or if you divide by a fraction. They get small if you multiply by a fraction or divide by a number bigger than 1. You may have thought that the estimate for the second example shown earlier, $87.3 \div 2.4$ was bit large at 50 when the true answer is 36.375. However, 87.3 is rounded up quite a bit to 100 and then it is divided by 2 that is rounded down quite a lot from 2.4. So a rounded-up number is being divided by a number that is rounded down, making it even larger. Consequently, we should expect an estimate that is quite a bit larger than the true answer.

Estimating Additions and Subtractions

Additions and subtractions are not always easy to estimate. Rounding to 1sf is sometimes helpful. For instance, $3.59 + 2.28 \cong 4 + 2 = 6$, so you would expect an answer of about 6. (The actual answer is 5.87.)

If the numbers you are adding or subtracting are very different in size, then the answer will be not much different to the larger number. For instance, if you are subtracting 0.287 from 15.3, the answer is going to be about 15. (The actual answer is 15.013.)

If you have to add up a list of numbers, there are some tricks that can help. If they are all about the same size, count up how many numbers there are, estimate an average value and multiply.

For example, suppose you want the sum of:

$$2.1 + 2.2 + 2.05 + 2.1 + 1.95 + 2.0 + 2.1 + 2.1 + 1.9 + 2.0$$

You could say that there are 10 numbers and they are all about 2.1 in size. Then the sum will approximately $10 \times 2.1 = 21$. (The actual answer is 20.5.)

If you need to add up several amounts of money, there is a similar way of estimating the total. If we are working in pounds sterling, add up the pounds, ignoring the pennies. Then count how many numbers there are and add half this number to your total.

For example:

$$£2.47 + £4.50 + £3.26 + £2.79 + £1.60$$

If you add the pounds, you get $2 + 4 + 3 + 2 + 1 = 12$. You have added 5 numbers and half of 5 is 2.5, so add £12 + £2.50 and your estimate is £14.50. (The actual answer is £14.62 .) This works for any decimal currency. It depends upon the pennies being any value randomly chosen between 0 and 99. If you are adding prices that are all set at just less than a pound, like £2.99, £4.99, etc. (as shops like to do), then you should just round up all the prices to a whole numbers of pounds.

Is It the Right Calculation?

Finally, whether you are working things out on paper or using a calculator or spreadsheet, there is the possibility that you are doing the wrong calculation in the first place. For instance, you have seen how repeated subtraction or repeated division can give you very different answers depending on the order in which you do the parts of the calculation. There are plenty of other ways in which you can formulate the wrong calculation to answer a given problem.

Is That a Reasonable Answer?

Always look critically at your answers to any problem and ask yourself "Is that reasonable? Is it what I would expect?" You might be thinking that this is stating the obvious. I can assure you that it isn't. I have seen so many completely crazy answers to examination questions.

That Was Silly!

Do not worry about making mistakes. I make them all the time. Everyone does. Just accept that you will get things wrong and look very hard at all your answers.

Practice Exercises

You can download practice examples of rounding to a given degree of accuracy, and of estimating answers, from http://numbersexplained.co.uk.

What Now?

I hope that Part 1 of this book has given you the understanding and procedures to use numbers confidently for everyday purposes.

I would also like to think that you might be curious about Part 2 which deals with more advanced number work. It contains mathematical ideas and techniques that are you are unlikely to need unless you intend to work in science or engineering but I hope that you might find them interesting for their own sake.

Stephen Miller

Birmingham, England, 2014

Summary

This is a list of the things that you should try to remember.

Counting	A set is a collection of things.
	The counting numbers are an ordered set of labels that we assign, in order, to every member of a set of objects. The last label that we use is the number of objects in that set.
Place Value	Every place value is worth ten times the one on its right. Zeroes are used to fill up empty place values.
Four Rules of Number	Addition is a movement up the number line and subtraction is an inverse movement the other way. Multiplication is repeated addition and division is the inverse – repeated subtraction.
	Addition (and subtraction): two evens or odds makes an even, even and odd makes odd. Multiplication (and division): both odds make odd, otherwise even.
	A number that will divide exactly into another one is called a 'factor'.
	A prime number can only be divided by 1 and itself.
BODMAS	Arithmetic expressions are not always worked out from left to right.
	Do anything in brackets first, then any divisions and multiplications, and finally any additions and subtractions.
Negative and Directed Numbers	Like signs are plus, unlike signs are minus.
	An even number of minus signs gives plus, and an odd number of minus signs give minus.
Fractions and Rational Numbers	A rational number is any pair of numerator and denominator. The numerator is the counting number and the denominator tells what scale is being used.
	To change a rational number into another equal one, multiply or divide both numerator and denominator by the same integer.
	A rational number is in its lowest terms if there are no common factors (other than 1) between the numerator and denominator.
	Rational numbers are continuous because we can make their denominators as large as we please. We can use this property to make measurements as accurately as we need.

Four Rules for Rational Numbers	To add or subtract rational numbers, convert them to a common denominator (unless they already have the same denominator). This is the denominator of the answer. Then add or subtract the numerators.
	To multiply two rational numbers, multiply the numerators and multiply the denominators. Then check that the answer is in its lowest terms.
	To divide by a rational number, multiply by its inverse. The inverse is the number 'turned upside-down'.
Mixed Numbers	Addition and subtraction of mixed numbers is done by adding or subtracting the integer and fraction parts separately and then combining the answers to give another mixed number.
	Multiplication and division of mixed numbers should always be done by converting to improper fractions. Then proceed as for any rational number multiplication or division.
Decimals	To change from a fraction to a decimal, divide the numerator by the denominator, adding as many zero decimal places as you need, until the answer terminates or recurs.
	To change a decimal into a fraction, take the place value of the last digit and make this the denominator. Then write the complete sequence of digits, without the decimal point or any leading zeroes, as the numerator.
Four Rules for Decimals	To add and subtract decimals, line up the decimal points.
	Multiplying or dividing by 10, 100, 1000, etc. involves moving the decimal point by the number of zeroes in the divisor. Move the decimal point to the right to multiply. Move the decimal point to the left to divide.
	When you multiply decimals, ignore the decimal points and multiply as if the numbers were integers. When you have the answer, count up how many decimal places there are altogether in the question and this will be the number of decimal places in the answer.
	To divide decimals, move the decimal point in the divisor so that its last digit is in the units place value. Move the decimal point the same number of places in the number you are dividing. Do the division. The decimal point in the answer will line up with the adjusted decimal point in the question.

Glossary

A   **accuracy**

How close the stated value of a **measurement** is to the true value. Either given as a number of **decimal places** (dp) or as a number of **significant figures** (sf).

addition

The procedure for increasing a number by another number. The **operator** for addition is the **plus** sign + and the result of an addition is called the **sum** or the **total**.

arithmetic

The rules that govern numbers and operators.

B   **base**

In a number system that uses **place value**, the base is how many each place is worth of the place on its right. In a **decimal** number, each column is worth 10 times as much as each column to its right, so that going to the left from the **decimal point** are units, tens, hundreds, thousands and so on, and going to the right are tenths, hundredths, thousandths and so on.

C   **calculation**

Working out the answer. More formally, the process of simplifying an arithmetic expression into a single number.

closed

A set of numbers is closed for a particular **operator** (such as addition) if any pair of numbers from the set always yields another member of the set when combined with that operator.

common

Common means 'belonging to both'. In fractions, a common factor is a number that will divide into both numerator and denominator. Common denominator means that two fractions have the same denominator (and so can be added or subtracted).

composite

A non-prime number.

continuous

The opposite of **discrete**. Continuous numbers such as **fractions** or **decimals** have the property that there are no gaps. No matter how close together two continuous numbers are, there are always more numbers in between them. Continuous numbers are used to **measure** things.

count, counting

A count is a description of a **set** of **discrete** objects. See **counting numbers** below.

| counting numbers | Counting numbers are an ordered set of labels. The labels are given to each member of a set. The last label used is the number of members in the set. Counting numbers usually refers to the set of **positive integers**, also called natural numbers. |

D **decimal**

Numbers that use base 10. The whole number columns are units, tens, hundreds, thousands and so on. The fraction columns are tenths, hundredths, thousandths, and so on.

decimal places

A measure of accuracy. It is how many place values are included. It is abbreviated to dp, for example 0.27 (2dp).

denominator

The bottom number of the pair in a **rational** number (a **fraction**). Each whole number is split into an equal number of parts shown by the denominator. Thus the denominator defines the scale used by the **numerator**.

discrete

Discrete objects cannot be split into smaller parts. Discrete **numbers** are used to count discrete objects. They go in steps with nothing in between, jumping from one number to the next.

difference

The result of a **subtraction**.

digit

In a **place value** number system like our decimal numbers, each symbol is a digit. For example, the number 125 has three digits. A **decimal** digit can be any of the ten symbols 0, 1, 2, 3, 4, 5, 6, 7, 8, or 9. The actual value of a digit depends upon its place value, so that in the number 125, the 1 is actually one hundred and the 2 is actually twenty.

directed number

A number with a **plus** or **minus** sign in front. If a number is shown without a sign, it is always taken to be **positive** – there is an unwritten plus sign in front.

division

Division is how many times a number can be repeatedly **subtracted** from another. It is the inverse of **multiplication**.

divisor

In a division, the divisor is the dividing number (not the number that is being divided).

E **estimation**

A very approximate (rough) answer to a calculation.

equation

A statement that two (or more) expressions are equal. It takes the form *some expression = some other expression*

equivalence set A fraction can be converted into another equal fraction by multiplying (or dividing) its numerator and denominator by the same integer. An equivalence set is an infinite set of fractions that are all equal to each other.

expression An **arithmetic** expression can be just a single **number** or it can be several numbers connected by **operators** such as +, - , etc. (It does not contain any **variables** – that would make it an algebraic expression.)

F **factor** The factors of an **integer** are all the integers that will divide into it exactly. If an integer has only 1 and itself as factors, it is a **prime number**.

fraction A number where the interval between each whole number has been broken into an equal number of parts so that you can count according to a new finer scale. It consists of a pair of integers written one above the other: $\frac{numerator}{denominator}$. The **denominator** gives you the scale: it tells you how many equal parts there are in the interval between each whole number. The **numerator** is then used to count using this scale.

finite A procedure or number that is not infinite. It might be very large, for example, in principle, you could count the number of grains of sand on a beach to yield an enormous but finite number.

I **identity** The member of a set of numbers that leaves a number unchanged after an **operation**. It is **zero** for **addition** and **subtraction** and it is 1 for **multiplication** and **division**. (The word has another meaning in algebra – it is a statement that is true for all values of the variables.)

improper fraction A **fraction** with a numerator greater than its denominator. Any number greater than 1 can be represented by an improper fraction. There are also negative improper fractions less than -1. See proper fractions.

inequation A statement that compares two or more **expressions**. It will use one of these five symbols: not equal ≠ , less than < , less than or equal ≤ , greater than > , greater than or equal ≥

	infinity	Infinity describes a procedure that goes on forever. For example, you could subtract **zero** from another number an infinite number of times. Hence, dividing by zero will not give an answer.
	integer	An integer is a **counting number**. The set of integers includes **zero**, the **positive** integers that increase infinitely from zero, and the **negative** integers that decrease infinitely from zero. Integers are **discrete** – they go in steps with nothing in between each number.
	inverse	The **operator** that reverses the effect of another. For example, the inverse of **plus** is **minus**. The word is also used for a particular pair of operations, e.g. +3 and -3. The multiplicative inverse of a number is called its **reciprocal** e.g. 3 and $\frac{1}{3}$ are reciprocals.
L	litre	The basic unit of capacity (or volume)
	lowest terms	A **fraction** whose **numerator** and **denominator** have no common **factors** except 1 is in its lowest terms. It is the member of its **equivalence set** that uses the smallest numbers.
K	kilogram	The basic unit of weight.
M	measure	A number that represents some continuous quantity such as a distance, a weight or a time. Unlike **counting**, measuring involves the consideration of **accuracy**.
	metre	The basic unit of length.
	minus	With two numbers, it means **subtract**. With a single number it means that the number is **negative**.
	mixed number	A number consisting of an integer and fraction added together. The addition sign is unwritten. An example is $2\frac{1}{2}$.
	multiplication, multiple	Multiplication is repeated **addition**. If you repeatedly add the same number, you get a regular pattern called the multiples of that number.
N	natural number	Natural numbers are the counting numbers, starting with 1 and increasing infinitely. They are the **positive integers**.
	negative	Less than zero.

	number	There are two main types of number. There are **discrete** counting numbers which are also called **integers**. There are also **continuous** numbers (with no gaps) such as **fractions** and **decimals** that are used for **measuring**.
	number line	A convenient way of thinking about numbers by spreading them out along a line so that they increase to the right of zero and decrease to its left
	numerator	The top number of the pair in a **rational** number (a **fraction**). It is the counting part of the number, and it counts along a scale defined by the **denominator** , not whole numbers.
O	**operator, operation**	An operator is an instruction to combine numbers. Examples covered in this book are **plus +**, **minus -** ,**multiply** x , and **divide** ÷. Division is also often shown by the fraction notation of a line /.
P	**percentage**	A number in hundredths. For an example in fractions, $\frac{1}{2} = \frac{50}{100} = 50\%$. As a decimal, move the decimal point in 0.5 two places to the right after the hundredths column = 50%.
	place value	The symbols that represent numbers are re-used by organising them into columns. The value in each column is called a **digit**. A digit is a fixed multiple of a digit to its right – that multiple is called the number **base**. The numbers we use in everyday life are **decimals** with a number base of ten. See **digit**.
	plus	With two numbers, it means **add**. With a single number it means that the number is **positive**.
	positive	Greater than **zero**.
	prime number	An **integer** that has only 1 and itself as **factors** so that it cannot be broken down into the **product** of two or more other numbers.
	product	The result of **multiplying**.
	proper fraction	A **fraction** with a **numerator** less than its **denominator**. All proper fractions are between 0 and 1. There are also **negative** proper fractions between 0 and -1.
Q	**quotient**	The result of a **division**.

| R | rational | The set of rational numbers is all the possible pairs of numerators and denominators. It includes all integers and fractions, both positive and negative. It is infinite in size and is closed for all four rules of number. |
| | real | The set of real numbers contains all possible sequences of decimal digits, both terminating and recurring, with the decimal point placed in all possible positions. It is infinite in size and is closed for all four rules of number. |

reciprocal

The multiplicative **inverse** of a number. For example, the inverse of 3 is $\frac{1}{3}$ because $3 \times \frac{1}{3} = 1$ which is the **identity** number for **multiplication**.

recurring

When a number is **divided** by another to give a **real** number answer, the result may be a decimal sequence of **digits** that repeats itself for ever. This is a recurring decimal (as opposed to a **terminating** decimal). To express a recurring number number exactly, the recurring sequence is written with a small dot over the beginning and end. For example, $\frac{1}{7} = 0.\dot{1}4285\dot{7}$ and $\frac{1}{3} = 0.\dot{3}$. In practice, recurring decimals are **rounded** to the required degree of **accuracy** e.g. $\frac{1}{7} = 0.143$ to 3sf.

rounding

If you have a very large number, say in millions, then you are probably not too concerned with the hundreds, tens and units values. So 36 255 432 would be rounded to 3 **significant figures**, giving 36 300 000 and may well be written as 36.3 million. Measurements are usually given to the nearest millimetre because that is the best that can be achieved in practice with tape measure, so a calculated value of 2.451̇6m will be rounded to 2.452m.

S **second**

The basic unit of time. Also a small unit of angle measurement.

set

A collection of objects.

	significant figures	A number of digits that are being used to express a required degree of accuracy. The number may be padded out with zeroes to show the place value of the significant figures e.g. 34 500 or 0.00328. However some zeros may be significant, so the accuracy is shown after the number e.g. 23 000 (3sf). In this example, the first zero is significant and the last two are not.
	subtraction	The procedure for reducing a number by another number. The **operator** for subtraction is the **minus sign** − . The result is called the **difference**. Subtraction is the **inverse** of **addition**.
	sum	The result of an **addition**. It is usually reserved for adding only two or three numbers. However, the word sum is often used to mean any **calculation**, not necessarily an addition.
T	**terminating**	When a number is **divided** by another to give a **real** number answer, the result may divide out exactly after a short sequence of **digits**. This is a terminating **decimal**. For example, $\frac{1}{4} = 0.25$.
	total	The result of an **addition**. It is usually reserved for adding a long list of numbers.
U	**unit**	Two meanings: 1) the **place value** where we start counting − the first column to the left of the decimal point or 2) the type of measurement e.g. metres for length, kilograms for weight, etc.
V	**variable**	In algebra, a variable is a member of a **set** (usually a set of **numbers**). It is usually represented by a letter of the alphabet. By convention, you should use letters such as i, j, k, l, m, n for **discrete** number variables and letters such as x, y, z for **continuous** number variables.
W	**whole number**	Another word for **integer**.
Z	**zero**	The starting point of all numbers because it is the result of **counting** an empty set of objects. It is the **identity** member for addition. Dividing by zero has no defined answer because you can **subtract** zero from a number an **infinite** number of times. Zero has an important function in filling up empty place values in large or small numbers.

Index